GASTRO-INTESTINAL PHYSIOLOGY

—the essentials

GASTRO-INTESTINAL PHYSIOLOGY
—*the essentials*

Thomas J. Sernka, Ph.D.
Associate Professor of Physiology
School of Medicine
Wright State University
Dayton, Ohio

Eugene D. Jacobson, M.D.
Professor of Physiology and Pathology
College of Medicine
University of Cincinnati
Cincinnati, Ohio

with a contribution by:
Tushar K. Chowdhury, Ph.D.
Professor of Physiology and Biophysics
College of Medicine
University of Oklahoma
Oklahoma City, Oklahoma

THE WILLIAMS & WILKINS COMPANY
Baltimore

Made in the United States of America

Library of Congress Cataloging in Publication Data

Sernka, Thomas J.
Gastrointestinal physiology, the essentials.

Includes index.
1. Alimentary canal. 2. Gastroenterology. I. Jacobson, Eugene D., joint author. II. Chowdhury, Tushar K., 1936- joint author. III. Title. [DNLM: 1. Gastrointestinal system—Physiology. WI102 S486g]
QP145.S45 612′.32 78-27862
ISBN 0-683-07720-1

Composed and printed at the
Waverly Press, Inc.
Mt. Royal and Guilford Aves.
Baltimore, Md 21202, U.S.A.

Dedication

The authors began this work as three colleagues, each of whom had a particular set of contributions to make. It was, therefore, a doubly sad loss to the two remaining authors when Dr. Tushar K. Chowdhury passed away after a brief and unexpected illness. He was greatly esteemed and his early loss from our collaborative text was not readily compensated for by us. His chapter on carbohydrate absorption in this book was the last complete writing of a respected friend and colleague. This book is dedicated to his memory.

Preface

This book is intended to serve as an introduction to gastrointestinal physiology for beginning medical and graduate students. Realizing that these students usually have but a week or so to devote to the subject, the authors have placed a premium on essentials and brevity. Physiological mechanisms have been stressed, while structural and biochemical details have been omitted. Wherever possible, the essential concepts have been highlighted by clear and simple illustrations. These essential concepts and key words have been recapitulated as brief summaries at the end of all chapters and as a comprehensive summary in the final chapter.

In recognition of the medical orientation of our readership, well-defined clinical examples of gastrointestinal pathophysiology have been added to all chapters.

The topics considered in this monograph include general gastrointestinal functions, secretion and absorption. Membrane transport and gastrointestinal circulation have been given special emphasis, because these subjects are basic to an understanding of gastrointestinal secretion and absorption.

Finally, the authors are grateful to the many individuals who helped in the technical preparation of this text. We especially appreciate the illustrations drawn by Molly G. Sernka. We also wish to recognize the critical advice of Dr. Michael Eade in the preparation of the chapter on the circulation.

Thomas J. Sernka
Eugene D. Jacobson

Contents

chapter 1

Introduction

the general functions of the gastrointestinal tract and the medical importance of this system

Gastrointestinal physiology is the study of the many normal functions of the organs that comprise the gastrointestinal tract. These include the esophagus, stomach, liver, pancreas, small intestine and colon. This portion of the body is ultimately responsible for the management of swallowed food and involves a number of processes essential for the conversion of food into a form that can be utilized by the rest of the body and for the elimination of waste material. In order to carry out these functions it is critical that some swallowed substances be moved as much as 450 cm from mouth to anus. It is also essential that various glandular structures in the gastrointestinal tract secrete juices which are added to ingested food to permit decomposition of complex molecules into smaller forms. Still another vital function of the system takes place in the small intestine where transport of small molecules from the lumen of the gut into the circulation takes place.

The processes of motility, secretion, absorption and excretion are regulated by both external and internal factors. The external factors include the autonomic nervous system and hormones. The internal factors involve certain properties of cells including excitability, membrane transport, biosynthesis and packaging of new molecules and release of synthesized materials. Gastrointestinal functions are also influenced by local tissue substances and nervous networks contained in the walls of hollow organs.

The gastrointestinal tract is an interface between the outside world and the body. Most materials from the environment which will enter the body, whether food or toxins, will do so by passing through the lining of the gastrointestinal tract. This system, therefore, serves as a protective barrier which excludes many items that have been swallowed into the hollow tubes of this system from gaining access to the rest of the body. Thus, for example, many bacteria are contained in the food that we swallow, yet these organisms are not permitted to reach the circulation as do digested food materials.

Other less well-understood functions of the gastrointestinal organs include their large contribution to the immunological defenses of the body, their active metabolism of many substances in the body and their contribution to the regulation of fluid and electrolyte balance.

More than any other organ system of the body the gastrointestinal tract consists of an heterogenous collection of dissimilar organs. The functions of the esophagus and the pancreas are as unlike one another as are the functions of the liver and the colon. This dissimilarity among the functions of the components of the system adds to our difficulty in understanding what gastrointestinal physiology is all about.

One may reasonably ask at this point: What is the medical importance of gastrointestinal physiology or even of the gastrointestinal tract itself? In the evolution of multicellular animals it became necessary to develop a specialized portion of the body to handle ingested food. The food had to be converted to a usable form and nonusable materials needed to be excreted. Thus, the economy of the body depends to a large extent on a smoothly functioning gastrointestinal system. Beyond the key physiological role of the gastrointestinal tract there is the medical significance of digestive diseases. In the United States the total economic cost of illness due to diseases of the gastrointestinal tract is exceeded only by the total cost of cardiovascular diseases and the cost of violent death and injury (accidents and crime). One-tenth of the total economic burden of all illness in this country is attributable to digestive diseases which account for one in seven admissions to general hospitals, one in four surgical operations, direct medical costs in excess of $10 billion and a total economic loss (direct medical costs plus costs of disability) in excess of $30 billion annually.

Among the very common diseases of the gastrointestinal system which confront most physicians are such entities as peptic ulcers, cirrhosis of the liver, hepatitis, gallstones, colitis, pancreatitis, diverticulitis, esophagitis, and cancers of the colon, pancreas, stomach, liver and esophagus.

Beside these serious and often life-threatening conditions there are a large number of symptoms referable to the gastrointestinal tract which cause people to seek the assistance of physicians. These include such common symptoms as indigestion, heartburn, vomiting, abdominal discomfort and pain, flatulence, constipation and diarrhea. In studies of otherwise healthy individuals in our society nearly one-half complain of the frequent occurrence of these symptoms. There is a flourishing over-the-counter business in American drugstores involving many remedies for these complaints.

Before one can hope to treat these symptoms or diseases in a rational manner it is essential that the normal functions of the gastrointestinal system be understood. At the present time there is a considerable body of information about gastrointestinal physiology and it is likely that the future will witness elucidation of many of the puzzling features of this complex set of organs.

Chapter 2

Mucosal Metabolism

how gastrointestinal cells work and divide

Two major functions of the gastrointestinal mucosa are to secrete and to absorb. Both secretion and absorption require metabolic work by the epithelial cells lining the gastrointestinal tract. To perform this work gastrointestinal cells utilize their own energy stores and that of the blood bathing them. These cells age rapidly and are sloughed into the lumen of the gastrointestinal tract. Their work of secretion or absorption is taken up by younger cells that have recently divided.

Bioenergetics

The metabolic work of secretion is oxidative. Enzymes that form the secreted product receive energy from oxidations of the tricarboxylic acid cycle. Most energy derives from oxidation of substrates circulating in the mucosal blood. These include glucose and free fatty acids. The blood flow to the secreting mucosa also provides the oxygen for substrate oxidation. Epithelial cell pO_2 is about 15 mm Hg and decreases with ischemia. Other substrates for oxidation include glycogen and triglycerides stored within the gastrointestinal cells. Mitochondria in gastrointestinal cells contain the oxidative enzymes that metabolize all the substrates from pyruvate and fatty acids into carbon dioxide. Other cytoplasmic enzymes metabolize stored glycogen and triglyceride into utilizable pyruvate and fatty acids.

Another oxidative pathway, the hexose monophosphate shunt, also forms pyruvate. In doing so, however, it synthesizes the NADPH essential to the formation of lipids. A bilayer of lipid is needed to form the unit membranes of gastrointestinal cells. Both plasma membranes and endoplasmic reticulum proliferate when secretion is stimulated. These membranes contain the transport enzymes for secretion.

The products of substrate oxidation in the gastrointestinal tract, as elsewhere, are reduced pyridine nucleotides, mainly NADH. These nucleotides, in turn, are oxidized within the mitochondria to yield the unit of chemical energy, ATP. The coupling of nucleotide oxidation to the phosphorylation of ADP provides the essential link between metabolism and secretion in the gastrointestinal tract.

A variety of substrate-specific ATPases in different gastrointestinal cells extract energy from the terminal phosphate bond of ATP to drive secretion. For example, $(H^+ + K^+)$-ATPase and HCO_3^--ATPase are thought to provide energy for H^+ and Cl^- secretion respectively by oxyntic cells.

Figure 2.1 is a summary of the metabolic steps important to secretion

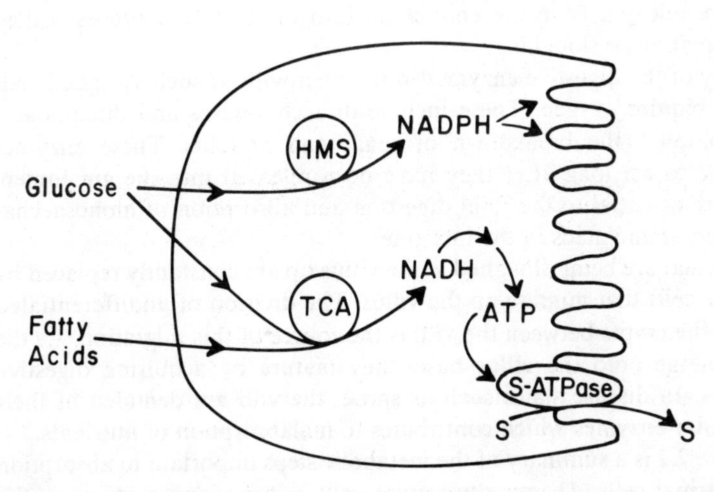

Figure 2.1. Basic metabolic mechanisms utilized in secretion of substance S by gastrointestinal cell. *TCA*, tricarboxylic acid cycle; *HMS*, hexose monophosphate shunt; *NAD(P)H*, reduced pyridine nucleotides; *S-ATPase*, substrate S-specific ATP phosphatase.

by gastrointestinal cells: 1) glucose and fatty acids derived mainly from the blood are oxidized; 2) reduced pyridine nucleotides are used to synthesize ATP or membranes; and 3) substrate-specific ATPase utilize ATP to effect secretory transport into the lumen.

The metabolic work of absorption is also oxidative. Glucose and fatty acids are oxidized to form reduced pyridine nucleotides, and the latter are oxidized with coupled phosphorylation of ADP. The intestinal basolateral membranes contain (Na^+ + K^+)-ATPase to drive absorption of Na^+ and accompanying anions.

Cytokinetics

Digestion is the enzymatic hydrolysis of carbohydrate, protein and fat. The digestive work that precedes absorption can be and often is hypoxic or even anaerobic. Hypoxic conditions are largely a consequence of the villus structure. Because the arteriolar and venular branches of the villus capillaries run countercurrent and closely adjacent to one another, physically dissolved oxygen in the incoming blood can diffuse across the base of the villus rather than travel its length in the capillary. This physiological shunting of oxygen from arteriolar to venular blood of the villus makes the villus tip hypoxic. The cells at the villus tip can survive only a limited time in such a hypoxic environment. After a day or two these hypoxic cells are released from the epithelium into the lumen, a process called desquamation or sloughing.

Many of the digestive enzymes in the microvilli of such sloughed cells do not require oxygen. These include disaccharidases and dipeptidases that complete the breakdown of sugar and proteins. These enzymes continue to act long after they have been released into the gut lumen. They are essential to the final digestion and absorption of monosaccharides and amino acids in the intestine.

Cells that are being sloughed at the villus tip are constantly replaced by younger cells that migrate up the villus. The division of undifferentiated cells in the crypts between the villi is the source of this migration. As the cells emerge onto the villus base, they mature by acquiring digestive enzymes. In disease states such as sprue, the villi are denuded of their microvillar enzymes which contributes to malabsorption of nutrients.

Figure 2.2 is a summary of the metabolic steps important to absorption by intestinal cells: 1) immature crypt cells migrate and mature on the villus; 2) digestive enzymes in the microvilli or brush border region act either in intact or sloughed cells; and 3) countercurrent blood flow insures hypoxia of the tip of the villus and continual sloughing.

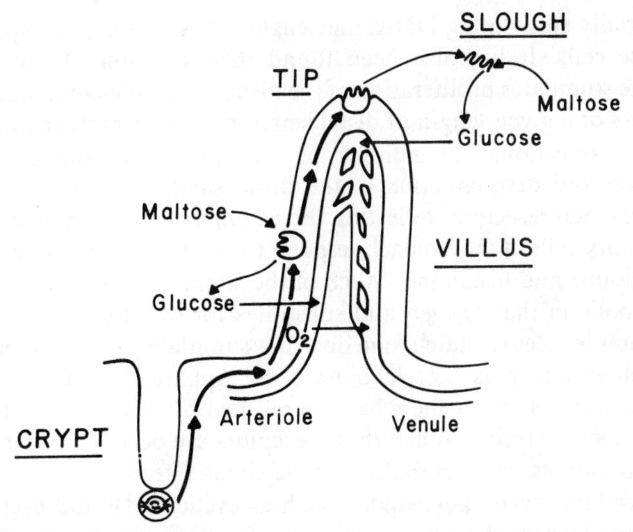

Figure 2.2. Metabolic aspects of absorption by gastrointestinal cells. Maltose is disaccharide of glucose.

Regulation

Normally, secretory glands line the pits in the epithelium and do not become hypoxic. Their arteriolar and venular capillary blood flows are separated by the invaginated epithelium. However, after severe hemorrhage there is a reduced blood volume (hypovolemia) which causes generalized mucosal ischemia and even secretory cells become hypoxic. Reduction in blood flow to the stomach drastically lowers intracellular pO_2 and ATP production and inhibits active transport of ions. Prolonged hypoxia of the secretory cells of the stomach leads to a type of stress ulcer.

Constant sloughing and cell renewal is characteristic of both secretory and absorptive epithelia in the gastrointestinal tract. The stimulus for epithelial cell division is unknown, although regulatory factors have been identified. The rate of cell sloughing at the villus tip exerts feedback control over the rate of cell division in the crypts. Thus, in sprue, where villus cells are attacked and destroyed at high rates, crypt cells multiply at correspondingly high rates to replace them. Another regulatory factor for epithelial cells growth appears to be the hormone, gastrin, which is synthesized by the so-called "G cells" of the gastric antrum. The trophic effect of this hormone is the proliferation of the gastrointestinal mucosa. Deprivation from food, as during total parenteral nutrition, leads to reduced production of gastrin and mucosal atrophy. Gastrin acts by sequentially stimulating DNA, messenger RNA and protein synthesis of sensitive cells. It has also been found that resection of the proximal intestine stimulates proliferation of the distal small intestinal mucosa. The thickness of a given length of distal small intestine greatly increases after proximal resection. The number of cells per villus and the rates of migration and desquamation in the distal small intestine also increase after proximal resection, reflecting the heightened cell turnover.

Secretory cells of the stomach respond to the three basic stimuli: gastrin, acetylcholine and histamine. Much of the stimulation by all three agents is metabolic in that oxygen and substrates for oxidation are consumed. For example, acetoacetate alone does not stimulate gastric secretory cells but acetoacetate plus acetylcholine does. Apparently, gastric secretory cells have receptors on their basolateral membranes for gastrin, acetylcholine and histamine. When these receptors are occupied by the appropriate stimulants, the metabolism of the cell is affected.

Intracellular cyclic nucleotides, such as cyclic AMP and cyclic GMP, have been identified as key mediators of cellular responses to external agents, such as hormones. The hormone activates a membrane-bound enzyme, adenylate cyclase, which catalyses the synthesis of cyclic AMP from intracellular ATP. The cyclic nucleotide then triggers a chain of reactions beginning with activation of a protein kinase, phosphorylation of an enzyme, calcium uptake and binding, activation of an ATPase and release of energy for secretion of solutes by mucosal cells. In intestinal diseases characterized by diarrhea due to secretion of fluid by the mucosa of the small intestine, for example, asiatic cholera, the preceding intracellular events occur. The stimulus for adenylate cyclase is an exotoxin of the bacteria which cause cholera. Other known stimuli include many of the prostaglandins, which are naturally occurring and which also cause diarrhea. The end result of stimulating the membrane-bound enzyme and causing accumulation of cyclic AMP inside intestinal mucosal cells is the active secretion of chloride from the tissue into the lumen of the gut. As solute is secreted, water is attracted osmotically and this hypersecretion of large volumes of fluid prompts diarrhea.

Summary

The metabolic energy for secretion and absorption in gastrointestinal mucosa comes from oxidation of glucose and fatty acids. NADPH is manufactured for lipid membrane synthesis and NADH for ATP synthesis. Dephosphorylation of ATP drives secretion of substances from the mucosal cell into the lumen by means of specific ATPases.

Mucosal cells are constantly being sloughed and replaced. In the intestine the tips of the villi are made hypoxic and prone to sloughing by diffusion of oxygen across the base of the villus. Digestive enzymes in the microvilli of epithelial cells remain active even after sloughing. Newly divided and undifferentiated cells migrate from the deep to the superficial mucosa to replace sloughed cells.

Cell growth and division in the gut is regulated by the rate of sloughing, the amount of functional mucosa and the trophic hormone gastrin. Gastric secretion is regulated by acetylcholine, histamine and gastrin, and diarrheal intestinal secretion is mediated by cyclic AMP.

Chapter 3

Membrane Transport

how gastrointestinal mucosa secretes and absorbs

Materials are transported across the gastrointestinal epithelial membranes in response to both electrochemical gradients (diffusion) and energy-dependent mechanisms (active transport). The rates of diffusional transport depend on the size and fat solubility of the solute for all substrates. Some substrates also have specific carriers which facilitate their transport rate. Some of these carrier-mediated substrates are transported against chemical and electrical gradients by the expenditure of metabolic energy.

Permeability

The rate at which dissolved substrates in blood or lumen can diffuse passively or permeate epithelial membranes of the gut depends on how small and how fat-soluble their molecules are. To illustrate the importance of molecular size, the membrane permeability to ethyleneglycol $[COH]_2$ is about 4 times that to mannitol $[COH]_6$, a molecule of the same polarity but three times as large. To show the even greater importance of fat solubility, the membrane permeability to diethylurea is about 3 times that to urea, a smaller but less fat-soluble molecule of the same basic structure. Diethylurea is about 4 times as fat-soluble as is urea.

To understand these permeability properties of gastrointestinal epithelia, we need to consider the structure and composition of the membrane.

As illustrated in Figure 3.1, membranes consist of a "fat sandwich." The outer layers adjacent to interstitial or cellular fluid are composed of porous protein that is not fat-soluble. Most water-soluble substrates can permeate these protein coats without hindrance. The middle layer between the protein layers, however, is itself composed of a bilayer of phospholipids that are very fat-soluble. Only substrates that are fat-soluble can permeate the fatty interior. Thus, for a molecule to permeate all the way from one side of a membrane to the other, it needs primarily to be fat-soluble enough to pass through the fatty membrane interior. This explains how a larger molecule like diethylurea can pass through membranes faster than a smaller but less fat-soluble molecule like urea.

Figure 3.1 also shows how small but fat-insoluble substrates can permeate membranes at all. The fat sandwich is occasionally perforated by a mosaic of tiny channels of porous protein that bring the interstitial and cellular fluids into continuity. If the substrate has a size smaller than these aqueous channels, it can permeate the entire width of the membrane directly. The protein channels permit the fatty interior to be bypassed. This explains how a small but fat-insoluble molecule like ethyleneglycol can readily permeate the membrane.

The principal driving force for the passive permeation of membranes

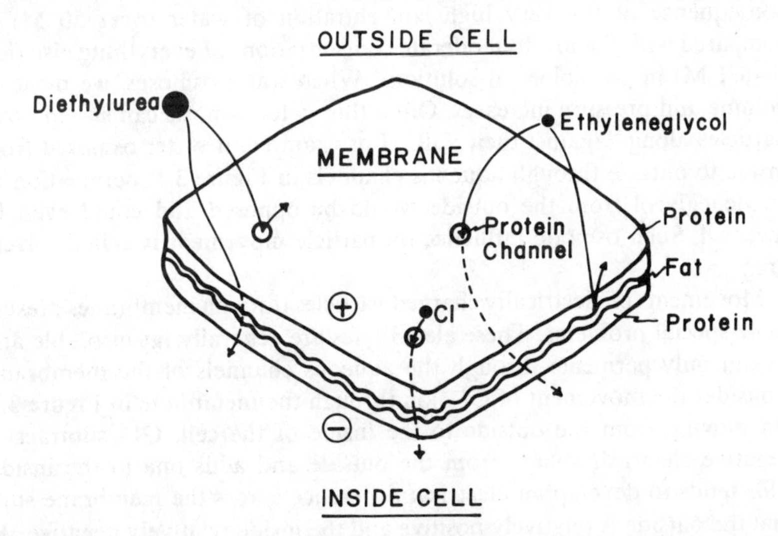

Figure 3.1. Passive permeability properties of gastrointestinal membrane. Large diethylurea molecule cannot pass through small channel but small ethyleneglycol molecule and Cl^- can. Fat-soluble diethylurea can dissolve through fatty interior but fat-insoluble ethyleneglycol and Cl^- cannot.

is diffusion. Diffusion is the net rate of particle movement in response to any concentration or electrical gradient. Permeation of particles through either the fat layer of the membrane or through the pores similarly occurs in response to any electrochemical gradient that exists across the membrane. The permeability of the membrane, depending on particle size and fat solubility, is expressed as a proportionality constant that relates rate of movement to the concentration difference. Thus, for example:

Rate of movement of diethylurea
(outside to inside of cell)
= permeability constant × (how fat-soluble) concentration gradient. (outside to inside of cell)

Net movement of water particles also occurs through membranes in response to a difference in water concentration on the two sides of the membrane. This diffusion process is called osmosis. Net water movement is from the dilute solution to the concentrated, that is, from the solution containing the lower concentration of dissolved solute particles (which is also the solution containing the larger number of water particles in a given volume). Osmosis is unique in bringing about volume movement of fluid without expenditure of energy at the membrane. Osmosis is a consequence of the very high concentration of water (over 50 M) as compared with the low to moderate concentrations of everything else (less than 1 M) in physiological solutions. When water osmoses, we measure volume and pressure increases. Often this water osmosis can sweep other particles along "against their will." For example, if water osmosed from inside to outside through aqueous channels in Figure 3.1, permeation of ethyleneglycol from the outside would be opposed and could even be reversed. Such osmotic influence on particle movement is called solvent drag.

Movements of electrically charged particles through membranes present some special problems. These electrolytes are generally fat-insoluble and so can only permeate through the aqueous channels of the membrane. Consider the movement of Cl^- ion through the membrane in Figure 3.1. On moving from the outside to the inside of the cell, Cl^- subtracts a negative electrical charge from the outside and adds one to the inside. This tends to develop an electrical difference across the membrane such that the outside is relatively positive and the inside relatively negative. As more and more Cl^- diffuses, the electrical charge grows larger. The electrical charge, it will be noted, opposes further diffusion of Cl^-, that is, the positive charge on the outside of the membrane attracts and the negative inside repels the negatively charged Cl^-. Finally, a balance will be achieved between the concentration difference that drove Cl^- into the cell and the electrical difference that opposes it. The balancing electrical difference, called the "Nernst" potential, is related to the concentration difference by a proportionality constant equal to 61 mV. Thus, for example:

$$\text{Balancing electrical difference (Nernst potential)} = 61 \text{ mV} \times \log \frac{(Cl^-) \text{ inside cell}}{(Cl^-) \text{ outside cell}}.$$

Notice that the logarithm of the concentration ratio is required for the final formulation. Remembering that the log of a ratio less than 1 is negative, it is evident that the inside of the cell will be negatively charged to the outside (blood) reference when the (Cl^-) is greater outside. When the outside Cl^- concentration is 10 times that inside, the balancing electrical difference will be −61 mV, since the log of 1/10 is −1. Knowing the concentrations inside and out, one can also calculate the balancing charge needed for a static situation. If this calculated charge turns out to be the same as the measured charge, then one may conclude that Cl^- movement occurred only in response to the chemical and electrical gradients across the membrane. This kind of movement is termed "passive transport."

Although the final passive distribution of ions across gastrointestinal membranes depends only on the prevailing chemical and electrical differences, the rate at which ions permeate depends on the size and charge of the protein channels. The size of aqueous channels ranges from a 15 Å diameter in the more permeable jejunum to only 4 Å in the less permeable colon. The "average" membrane in the body has channels with a diameter of 8 Å. The net charge of protein channels depends on whether the number of anionic carboxyl or cationic amino groups predominates in the channel protein. In the gallbladder the anionic carboxyl groups outnumber the others, and the protein channel has a net negative charge. This makes the gallbladder epithelium more permeable to cations, because the opposite charges of permeating cation and channel protein attract one another. In the stomach, on the other hand, the cationic amino groups predominate in the channel protein and hence gastric mucosa is more permeable to anions.

A contribution to the electrical difference across membranes is provided by the net charges of protein that may be present on one side but not the other of a membrane. For example, albumin is a negatively charged protein found in much higher concentration in the plasma than in interstitial fluid, and intracellular, negatively charged protein is likewise absent from the interstitial fluid. Because these three fluid compartments must maintain electroneutrality of their respective solutions, permeable cations like Na^+ or K^+ will tend to accumulate in the plasma or cell and permeable anions like Cl^- in the interstitial fluid. When the concentration difference equals the opposing electrical difference, net movement of ions ceases and "Donnan" equilibrium is established. Quantitatively, this process concentrates cations and depletes anions by only about 5% in the protein-containing compartment. Such a small concentration difference, in turn, will be balanced by an equally small Nernst potential, about 1.4 mV (interstitial fluid positive). Since the actual electrical difference from interstitial fluid to the inside of an intestinal cell is about 20 times as large, one may appreciate that the contribution of protein inside gastrointestinal cells to their potentials is only a very minor one.

Some solutes can exist in both charged (ionized) and uncharged forms. These weak electrolytes are variably fat-soluble according to the fraction in the uncharged form. The fat solubility of weak electrolytes depends on the acid concentration. For a weak acid like aspirin, a high acid concentration produces more of the uncharged, fat-soluble form:

$$\text{Aspirin—COO}^{\ominus} + H^+ \rightleftharpoons \underset{\text{(uncharged, fat-soluble)}}{\text{Aspirin—COOH.}}$$

For a weak base like morphine, the opposite holds:

$$\text{Morphine-NH}_3 + H^+ \rightleftharpoons \underset{\text{(charged, fat-insoluble)}}{\text{Morphine—NH}_4^+.}$$

This becomes an important consideration where the absorption of drugs from the stomach is concerned. The stomach produces high concentrations of acid that converts, by mass action, aspirin to the uncharged and morphine to the charged form. Since only the uncharged form is fat-soluble, aspirin can be absorbed from the stomach but morphine cannot (both drug molecules are too large to permeate the protein channels). Morphine trapping in the gastric juice is used to diagnose morphine addiction.

In gastrointestinal mucosal membranes, characteristically, concentration differences that affect solute permeation are determined by the microenvironment immediately adjacent to the membrane. Due to the extensive folding of these membranes the fluid layer adjacent to the membrane is not identical to the fluid in the middle of the lumen. No amount of peristaltic mixing can affect this unstirred layer. The unstirred layer in the intestine has two important consequences. First, it comprises the principal diffusion barrier in the absorption of fats, which easily permeate intestinal membranes themselves. Second, due to the presence of acidic groups in the glycocalyx of microvilli that occupy this region, the unstirred layer has a lower pH than that of the lumen. This second factor influences the absorption of weak electrolytes, including many drugs.

Finally, consideration must be given to permeability paths between epithelial cells of the gastrointestinal tract. These intercellular or paracellular shunt pathways are mainly of importance to the permeation of small electrolytes, because the tight junctions between epithelial cells prevent transfers of larger solutes like protein. The protein cement of the tight junctions, however, serves as alternate aqueous channels for passage of ions and water. In the ileum, for example, about 4 times as much fluid and electrolyte traverses the paracellular shunt as crosses the intestinal membranes. The gallbladder is even a "looser" epithelium having a shunt 20 times the membrane permeability. The fundic gastric and colonic mucosae, on the other hand, are "tighter" epithelia having shunts equivalent to only one-fifth the membrane permeability. Generally speaking, those epithelia responsible for absorption have large intercellular shunts and large membrane protein channels, and those epithelia responsible for storage or secretion have small shunts and small channels.

Figure 3.2 is a summary of the factors that determine the relative permeability of a gastrointestinal epithelium to a given solute: 1) the size and fat solubility of the particle, 2) the chemical and electrical difference across the membrane to the unstirred layer, 3) the proportion of uncharged to charged form of the particle, and 4) the proportion of intercellular to transcellular permeability.

Transport

All of the considerations involving permeability of membranes that we have discussed thus far are included as "simple" passive transport. An-

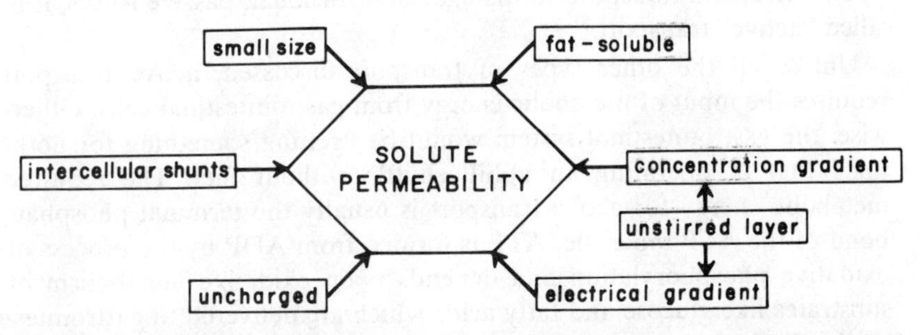

Figure 3.2. Positive influences on solute permeability in gastrointestinal epithelial membranes.

other class of transport is distinguished and characterized by the interaction of the permeating substance with the mosaic protein of the membrane. This interaction has all the same characteristics as that between substrate and enzyme; *i.e.,* the reaction shows saturation at high substrate concentration, competition between structurally similar substrates, specificity to stereoisomers of substrate and inhibition by metabolic poisons. For these reasons, the "carrier" protein that binds and carries substrate from one side to the other of the membrane may be considered a specialized membrane enzyme. Only the names have been changed from those used in Michaelis-Menten enzyme kinetics. Thus we speak of the "transport maximum" (T_{max}) instead of V_{max}. This class of transport that requires the participation or mediation of carrier protein in membranes is called "mediated" transport.

Mediated transport serves the needs of the gastrointestinal cell in several ways. First, it can allow for larger, water-soluble substances to cross the membrane. For example, the sugar fructose is somewhat too large to permeate the porous protein channels of the intestinal microvillous membrane. By combining with a specific membrane carrier, however, the fructose-protein complex can diffuse or be shuttled from the lumen to the cell interior. Because this type of mediated transport facilitates the otherwise slow passive transport of fructose, it is called "facilitated" diffusion (or transport). Second, mediated transport can allow essential substrates to move against the prevailing diffusional forces of the chemical and electrical differences in order to be secreted or absorbed. For example, H^+ ion can be removed from relatively neutral cytoplasm of gastric oxyntic cells into the highly acidic contents of the gastric lumen against a million-fold concentration gradient. Similarly, Na^+ ion can be removed from relatively low concentrations in intestinal chyme and be absorbed into plasma containing much higher Na^+ concentrations. Because this type of mediated transport occurs against diffusional, passive forces, it is called "active" transport.

Unlike all the other types of transport discussed, active transport requires the input of metabolic energy from gastrointestinal cells. Otherwise, the gastrointestinal system would be "getting something for nothing," that is, operating an uphill shuttle without fuel. The required metabolic energy for active transport is usually the terminal phosphate bond of the ATP molecule. ATP is formed from ADP by the process of oxidative phosphorylation and depends upon oxidative metabolism of substrates like glucose and fatty acids which are delivered to gastrointestinal cells by the circulation. Metabolic utilization of ATP to provide energy for active transport depends upon an enzyme, which hydrolyses

ATP, namely an ATPase, that is probably identical with the carrier protein in the membrane. Each active transport process is thus driven by the activity of membrane ATPase. Often the ATPase activity is specifically stimulated by the actively transported substance. For example, a gastric ATPase is stimulated by H^+ ion and an intestinal ATPase by Na^+ ion.

The membrane ATPase that drives the active transport of a given substance does not necessarily have to be specific for that substance. Active transport of sugars and amino acids in the intestinal mucosa lack specific ATPases, for example. Their active transfers come about by specific coupling or linkage to the active transport of Na^+ at this site. Thus, it is the Na^+-specific intestinal ATPase that accounts for the active transport of sugars and amino acids. The linkage or coupling for these transport processes is a carrier protein having sites for both the sugar (or amino acid) and for Na^+. Furthermore, the binding of Na^+ exerts a favorable allosteric effect on the binding of the other substance, say sugar or amino acid, and *vice versa.* The binding of carrier to Na^+ and sugar or amino acid occurs on the luminal membrane of the intestinal mucosal cell, and the complex of carrier protein and attached substances then diffuses to the intracellular side of the membrane, where the Na^+ and sugar or amino acid is released. The transfer across the luminal membrane is carrier-mediated but does not require metabolic energy—hence, it is facilitated diffusion. This process is shown graphically in Figure 3.3.

The metabolic energy input in coupled transport occurs at the opposite, basolateral membrane of the intestinal mucosal cell. Here there is another carrier protein with a binding site for Na^+ but not for sugar or amino acid. After binding Na^+ this carrier acts as an ATPase to receive metabolic energy, so that the Na^+ transfer can occur against a chemical and electrical gradient. The Na^+ released by this carrier is removed from a low concentration and electrically negative cytoplasm into a high concentration and electrically positive interstitial fluid on the blood side. This transfer across the basolateral membrane is carrier-mediated and does require metabolic energy—hence, it is active transport (Figure 3.3).

The sugar within the intestinal cell does not have a carrier at the basolateral membrane and must diffuse out into the blood down its concentration gradient by simple passive transport. Basolateral membranes in general are much more permeable than luminal membranes to water-soluble substances.

Coupled transport results in the active uptake of not only Na^+ ion but also sugar or amino acid, because their entry steps are linked. As more Na^+ is actively expelled at the basolateral membrane, more Na^+ and sugar are drawn into the cell by the diffusion gradient that is created.

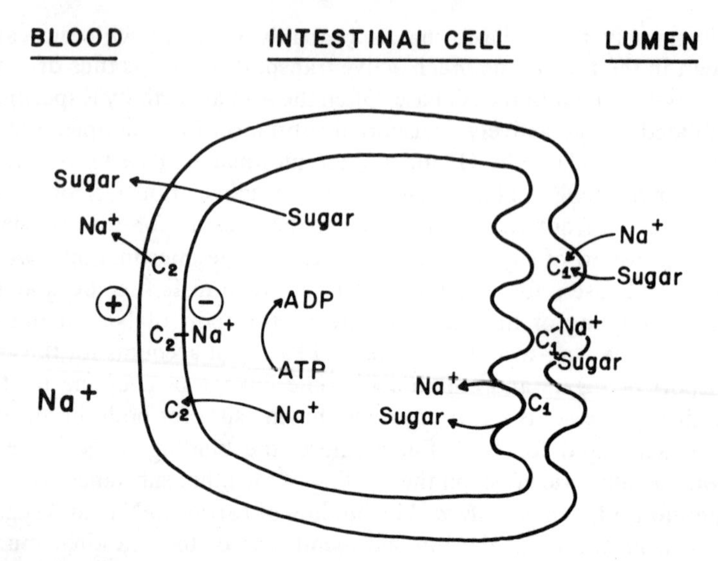

Figure 3.3. Scheme of the essential steps in the coupled transport of Na^+ and sugar across the intestinal mucosal cell. C_1 and C_2 are carrier proteins in the luminal and basolateral membranes, respectively.

Other transport processes also can be coupled. Thus, the pH gradient that is maintained by the active transport of H^+ ion by gastric oxyntic cells can initiate the absorption of a weak acid like the drug aspirin. In this case the accelerated entry of the weak acid occurs as a result of enhanced solubility in membrane lipid rather than by way of a carrier with two sites.

The different types of transport utilized in the secretion and absorption of various substances in the gastrointestinal tract are summarized in Table 3.1. Note that most classes of substances are transported in more than one way. Sugars, for example, can be actively transported in the absence of sodium or coupled to Na^+ transport. In coupled transport, transport at the luminal membrane is by facilitated diffusion and at the basolateral exit by simple passive transport. Virtually all substances enter or leave the gastrointestinal tract partly or wholly by simple passive transport. For this reason the simple passive classification is all-inclusive. Some substances can only be secreted or absorbed by simple passive transport (* in Table 3.1). These include water, fats and drugs. Water is always secreted or absorbed in response to its concentration or osmotic gradient, which is usually created by the active transport of electrolytes present at much lower concentrations. Not only fatty acids and glycerides but also fat-

Table 3.1
Classification of transport in the gastrointestinal tract

SIMPLE PASSIVE	FACILITATED PASSIVE	ACTIVE	COUPLED ACTIVE	PINOCYTOSIS
Water*	Sugars	Electrolytes	Electolytes	Proteins
Electrolytes	Amino Acids	Sugars	Sugars	
Fats*	Electrolytes	Amino Acids	Amino Acids	
Drugs*		Dipeptides	Vitamins	
Vitamins		Vitamins		
Sterols		Bile Acids		
Sugars				
Amino Acids				

* Substances transported only by simple passive transport.

soluble vitamins and cholesterol are absorbed by simple passive transport into the gastrointestinal mucosa. Drugs that exist as weak electrolytes can be passively absorbed in their undissociated forms. Finally, minute amounts of undigested protein can be absorbed intact into intestinal cells by the process of pinocytosis, or membrane engulfment of substances with solution. This process may be important in the absorption of maternal antibodies from milk consumed by the newborn and in the acquisition of certain allergies in the adult. Many proteins are secreted as enzymes throughout the gastrointestinal tract in a reverse pinocytosis, or exocytosis. Since exocytosis is more specific and requires more metabolic energy input, protein secretion has been classified as active transport.

Disordered Transport

A clinical condition involving disorganization of membrane transport occurs as a result of damage from topical contact with the gastric mucosa induced by such commonly used agents as aspirin and ethyl alcohol.

Normally there is a slow back-diffusion of some of the secreted H^+ in gastric juice, since the gastric mucosal membrane is not very permeable to cations. The mucosal cell actively transports Na^+ from inside to out across the basolateral membrane and actively transports Cl^- across the luminal membrane into the juice. Aspirin is a weak acid. Contact with the mucosa after swallowing the drug results in rapid diffusion of aspirin from the more acidic juice into the cell where the higher pH favors dissociation of aspirin with liberation of H^+. Lowering intracellular H^+ concentration adversely affects enzymatic processes which provide energy for active transport. The cell becomes unable to extrude either Na^+ or Cl^- and therc are changes in the charge of the membrane pores. The membrane becomes much more permeable to the diffusion of cations through the pores and Na^+ appears in the juice. More importantly, H^+ floods into the cell from the acidic juice, further impairing cell metabolism, triggering release of proteolytic enzymes from the mucosa and causing spasm of microcirculatory vessels. This clinical condition is termed "gastritis" and may be the cause of severe hemorrhage

Summary

The permeability of the gastrointestinal mucosa governs the passive transport of substances into and out of the lumen. Substances may permeate largely between cells in loose epithelia like the intestinal mucosa or through cells in tight epithelia like gastric mucosa. The force driving passive transport of solutes is the electrochemical gradient across the gastrointestinal mucosa. The force driving passive transport of water is the osmotic gradient across the gastrointestinal mucosa. Lipid solubility of solutes speeds up their passive transport through gastrointestinal membranes.

Substances are also actively transported against their electrochemical gradients into and out of the lumen. This movement requires attachment of the substance to a membrane carrier protein with the expenditure of metabolic energy, generally the hydrolysis of ATP. The carrier ATPase binds not only to the primary solute transported but also to other secondary solutes. These latter solutes are moved against their electrochemical gradients by coupling to primary active transport. Aspirin damage to gastrointestinal mucosa begins with inhibition of active ion transport.

Chapter 4

Gastrointestinal Circulation

how blood flow is regulated and serves the functions of the gastrointestinal organs

Among the many systemic regional circulations in the body, the gastrointestinal is the largest in terms of the proportion of cardiac output it receives, namely 30%. Its two major arteries, the celiac and the superior mesenteric, each carry a larger volume flow per minute than any other branches of the aorta. This circulation is complex, both structurally and functionally, and relatively little is known about its physiology. Perhaps as a consequence, disorders of the human gastrointestinal circulation are not easily diagnosed or treated and often prove fatal.

Microcirculation

The gastrointestinal circulation consists of the separate blood vessels of the salivary glands, pharynx and esophagus, stomach, pancreas, liver, small intestine and colon. The amount of information about each of these separate circulations varies considerably. The largest portion of the entire gastrointestinal circulation is located in the abdomen and is often referred to as the "splanchnic circulation," because the splanchnic nerves course along with the arteries of the region. The splanchnic circulation supplies the stomach, pancreas, small intestine, colon, liver, spleen, omentum and mesenteric tissues. The term "mesenteric circulation" is often used to designate the intestinal circulation.

The splanchnic circulation has two complicating features, one anatomic and the other functional. The structural complexity is caused by the circulation being arranged both in parallel and in series. Thus, the hepatic branch of the celiac artery passes directly to the liver. The hepatic artery is in parallel with the other branches of the celiac artery and with the mesenteric arteries. The splenic, left gastric and mesenteric arteries are distributed to the many organs of the region, from which the blood drains to the large portal vein connecting the capillaries to these organs with the capillaries of the liver itself. In the liver portal venous blood mixes with the hepatic arterial blood. This mixed blood pours into the hepatic veins en route to the heart. This arrangement is depicted schematically in Figure 4.1.

The second impediment to any simple description of the hemodynamics of the splanchnic circulation is caused by the diversity of organs whose blood supply is derived from these vessels. The celiac artery, for example, provides blood to the following functionally dissimilar organs: stomach, spleen, liver, pancreas and duodenum. Since changing levels of activity of an organ can cause a change in its blood flow, it would be difficult to determine the cause of an increase in celiac artery blood flow after consumption of a meal, a process which affects the functions of these many organs in different ways.

In the stomach, small intestine and colon the wall of each hollow organ is composed of different layers of tissue with different functions, and the circulation of each of these organs is arranged in parallel-coupled vascular circuits, as depicted in Figure 4.2. The microscopic vessels of hollow organs are also composed of segments arranged both in parallel and in series. Vascularization of the inner lining of these organs, the mucosa, is more dense than in the submucosa or muscular layer of the wall. The greater population of mucosal blood vessels coincides with the greater metabolic activity of this layer.

When the arteries to the stomach or bowel penetrate the wall of the organ from the outside, the diameter of vessels progressively diminishes. At a diameter of less than 25 μ the vessel is termed an "arteriole" and is characterized by a relatively thick smooth muscle wall relative to its internal diameter. The wall is highly responsive to stimuli, and the total cross-sectional area of all arteriolar branches of an artery is enormous; consequently most of the resistance to the flow of blood through the gastrointestinal arteries occurs at the level of arterioles, leading to the designation of this portion of the microcirculation as the "resistance vessels." Changes in the internal diameter of arterioles have the largest effect on the rate of blood flow through the entire organ.

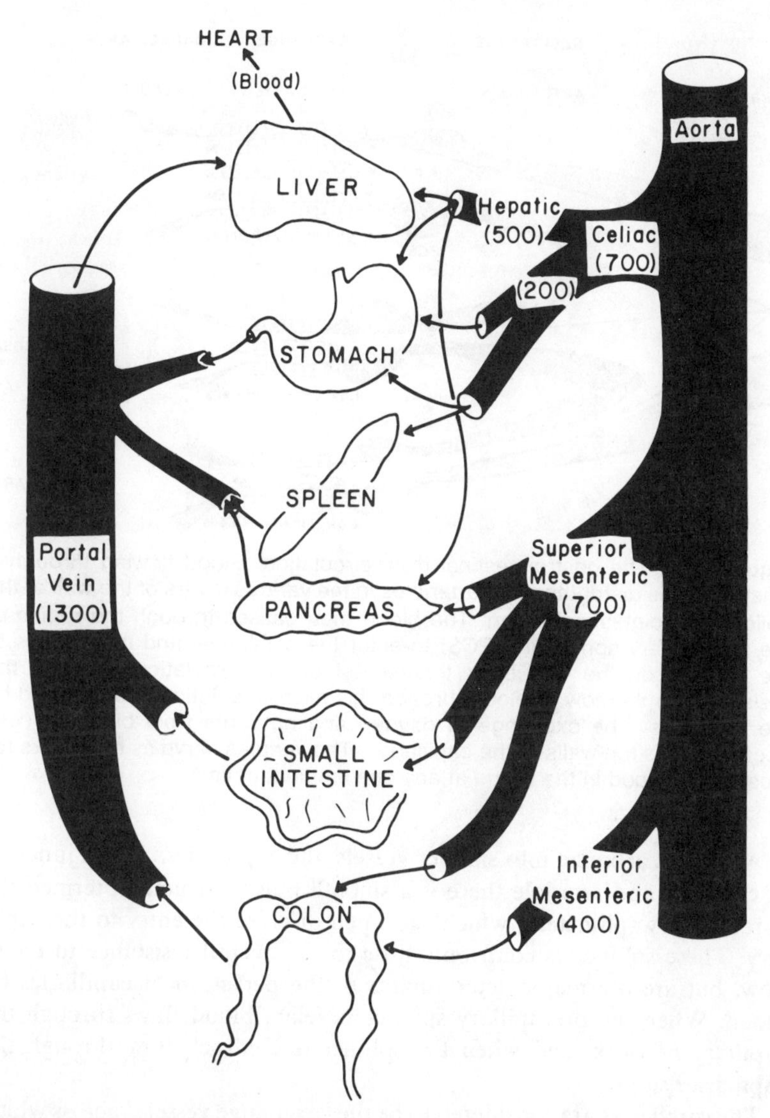

Figure 4.1. Organization of blood flow to and from splanchnic organs. For most splanchnic organs blood flows to each through a branch of one of the three major arteries from the aorta and then collects into the portal vein to drain to the liver. The liver also receives blood from the hepatic arterial branch of the celiac artery. Hence, blood passes both in parallel circuits and in series circuits through the splanchnic vessels. Numbers in parentheses indicate the blood flow in milliliters per minute through the named vessels that might be expected in an average adult male.

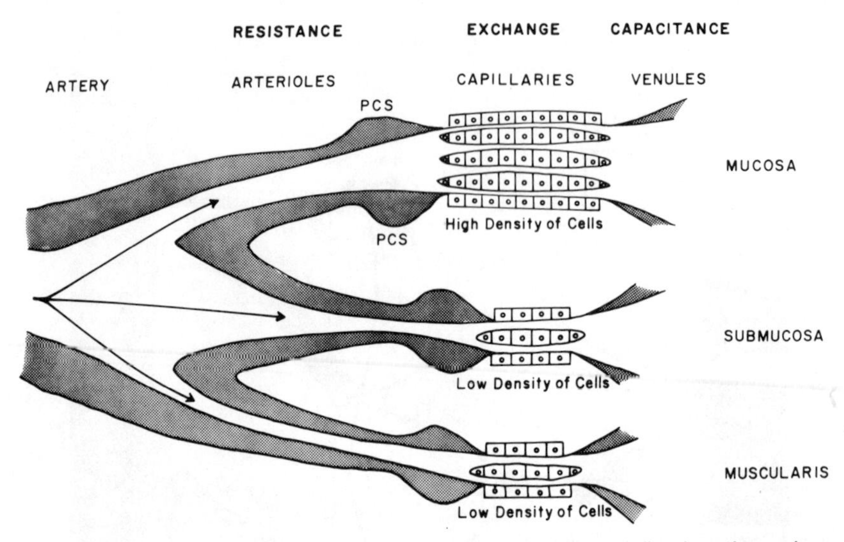

Figure 4.2. The gastrointestinal microcirculation. Blood flowing through a small artery is distributed into arterioles in the various layers of the wall of the hollow gastrointestinal organ. The blood then passes through the region of the precapillary sphincters (*PCS*) to enter the capillaries and then drains to the venules on the way back to the rest of the circulation. Most of the resistance to the flow of blood through this microcirculation is developed by the arterioles. The exchange of oxygen and nutrients from blood to cells occurs across the walls of the capillaries. The venules serve as reservoirs for most of the blood in the tissue at any one moment in time.

Arterioles arborize into smaller vessels, the capillaries. At the junction of capillary and arteriole there is a smooth muscle structure, termed the "precapillary sphincter," which can open or close the entry to the capillary. These sphincters contribute little to the overall resistance to blood flow, but are the major determinants of the perfusion of capillaries by blood. When the precapillary sphincters relax, blood flows through the capillary network, and when the sphincters contract, flow through the capillaries stops.

The capillaries are considered to be the "exchange vessels" across which nearly all transfers occur between intra- and extravascular compartments: transfers of fluid, electrolytes, heat, nutrients, metabolites and gases. The capillary wall has little or no smooth muscle except at its origin where the sphincter regulates its perfusion.

Distal to the capillaries are the venules and veins. This portion of the gastrointestinal microcirculation serves two functions. When its smooth muscle is stimulated and contracts, there will be a small increase in overall resistance to the flow of blood through the organ which will be reflected back into the capillaries. Such transmitted back-pressure increases mean hydrostatic capillary pressure, thereby increasing the rate of movement of fluid across the capillary wall. In addition, the venules and veins contain most of the blood in the circulation of gastrointestinal organs and act as "capacitance vessels" for the rest of the body. Within 1 minute after exercise is initiated splanchnic venules and veins experience an increase in smooth muscle tone and contribute 1 liter of blood from their reservoirs to the rest of the circulation for distribution to active muscles in the human body.

Normally blood passes from artery to arteriole to capillary to venule to vein. It is also possible for some shunting of blood from the arterial portion of the circulation to the venous which may occur in one of two ways. There are anatomically demonstrable arteriovenous shunts in gastrointestinal organs, but these constitute only a small fraction of the total population of vessels. In advanced cirrhosis there are numerous open shunts in the liver which reduce blood flow to the capillaries and thereby reduce delivery of oxygen, substrates and substances to be metabolized by the hepatocytes. A more important normal phenomenon is physiological shunting, which consists of a redistribution of blood flow away from one capillary bed into another (Figure 4.2). If this redistribution causes blood to flow away from areas where there is a high density of cells (mucosa) into areas where there is a low cell density (submucosa), the result will be a decreased extraction of oxygen from the blood as it passes through the tissue, corresponding to the effects of an actual anatomical arteriovenous shunt. Redistribution of blood flow away from the mucosa into the muscularis occurs in response to sympathetic nervous stimulation.

Measurement of Blood Flow

In patients and volunteers it is possible to measure splanchnic blood flow indirectly. Fairly accurate estimation of blood flow through an organ can be made if the arterial blood carries a measurable marker such as a dissolved dye and if the dye is either extracted by the organ into a secreted juice or passes through the capillaries into the venous blood. In patients a commonly used marker is indocyanine green dye. It is necessary to measure the concentration of this dye in the blood entering the liver and in the blood leaving the organ. This requires introduction of a needle into the brachial artery to sample blood that is representative of all arterial blood, including hepatic artery blood. An intravascular catheter is also advanced into the hepatic vein by way of a femoral vein and the inferior vena cava. It is necessary to infuse the dye at a rate equal to the rate at which the liver extracts the dye from the blood into the bile. Since the liver is the only organ which extracts the dye, the rate of infusion of dye intravenously is simply the rate which keeps concentration of the dye constant in the arterial blood. The amount of dye entering the liver (arterial concentration × arterial blood flow) equals the amount leaving the liver (venous concentration × venous blood flow plus the amount of dye being secreted into the bile). Since all of the blood which flows through the splanchnic organs also flows through the liver, hepatic blood flow measurement is a determination of splanchnic blood flow.

With these considerations in mind the calculation of splanchnic blood flow is derived from the rate of infusion of dye intravenously required to keep the arterial concentration at a steady value divided by the difference between the concentration of the dye in the arterial blood and in the venous blood draining from the liver. The numerator of this equation is expressed in grams of dye infused per minute, the denominator in grams of dye per milliliter of plasma and the quotient is equal to milliliters per minute (a flow value). Knowing the hematocrit of blood (which is about 50, that is, 50% of the blood is composed of red cells and 50% of plasma) one can correct for the plasma flow value to convert it into a blood flow value. In adult males the resting splanchnic blood flow estimated in this manner is about 25 ml/min/kg body weight which is close to 2 liters/min. Splanchnic flow serves organs weighing about 4 kg.

There are other techniques for obtaining indirect measurements of portions of splanchnic blood flow. An indicator dilution method has been developed to measure superior mesenteric artery blood flow in conscious human subjects. The method requires catheterization of the superior mesenteric artery and of the corresponding vein (via catheterization of the umbilical vein).

A clearance technic employing trace amounts of radiolabeled aminopyrine has been developed to measure gastric mucosal blood flow in intubated conscious human subjects. Aminopyrine is a poorly dissociated base in the blood and readily crosses the lipid membrane of the gastric mucosa to reach the gastric juice. At the low pH of the juice aminopyrine dissociates readily and behaves like an ionized solute, thereby becoming unable to recross the membrane back into the blood. Thus, aminopyrine is trapped in the gastric juice and becomes highly concentrated. The major limiting step in this accumulation of aminopyrine in the juice is the rate of delivery of the drug by the mucosal circulation. Consequently, the amount of aminopyrine in the juice reflects the mucosal blood flow which can be calculated as:

$$\text{Mucosal blood flow} = \frac{\text{Gastric secretory rate} \times \text{aminopyrine concentration in gastric juice.}}{\text{Aminopyrine concentration in blood}}$$

The preceding techniques represent indirect approaches to measuring blood flow to gastrointestinal organs. The major disadvantages of these techniques for use in diagnosing disease is that they are technically difficult to perform and require a steady state to obtain adequate measurements. In patients suffering catastrophic low flow states of major splanchnic vessels, it is possible to diagnose the ischemia or to identify a point of vascular occlusion in advance of surgery. The technique widely in use for this diagnostic purpose is termed "angiography" and involves advancing an intravascular catheter up the aorta from the femoral artery to the vicinity of the occluded or ischemic vessel (usually the superior or inferior mesenteric artery) to permit the rapid injection of a radiopaque fluid. Rapid sequential x-ray films reveal failure of the opaque solution to flow through the occluded or constricted artery at normal rates. In addition, the angiographer can infuse the opaque solution at a rate which barely exceeds the normal rate of blood flow for a fraction of a second to induce reflux of the opaque fluid and thereby determine the rate of blood flow.

The gastrointestinal organs are well-perfused relative to the rest of the body. At rest the mucosal lining of a hollow organ (such as the stomach or gut) and the metabolically active solid organs (pancreas or liver) receive a blood flow of about 1 ml/min/g of tissue. When there is an increase in the activity of these tissues, as might occur during brisk secretion following a meal, this level of flow will be increased several-fold. Lower rates of tissue perfusion are found in the muscular outer coat of the hollow organs where flow is only a few tenths of a milliliter per minute per gram of tissue.

Each visible branch of a major splanchnic artery courses over the serosal surface of the hollow organs and gives rise to smaller branches which pass through the serosa and muscularis and terminate in an extensive submucosal network of small arteries. From this arborization the arterioles arise and carry blood into the dense mucosal capillary vessels. The submucosal vascular network assures anastomotic connection of the circulation between adjacent segments of the hollow organs. This collateralization protects somewhat against ischemia of the organ in diseases which reduce the arterial supply.

CONTROL OF THE CIRCULATION

General Hemodynamic Factors

Blood flow to any gastrointestinal organ is the vector of several factors which operate simultaneously. Among these are general hemodynamic factors, such as systemic arterial pressure, cardiac output and blood volume. A sizable reduction in any of these general circulatory factors is translated into a decrease in gastrointestinal blood flow.

Autonomic Nervous System

The autonomic nervous system has a significant influence upon blood flow to the digestive organs. Activation of parasympathetic nerves increases both secretory and motor activities of gastrointestinal organs, thereby enhancing tissue metabolism, oxygen consumption and generation of metabolites which relax arteriolar smooth muscle. Vasodilation results and blood flow increases. Parasympathetic activation may also cause visceral smooth muscle to contract vigorously, and pressures may be generated in the wall of the hollow organ which exceed blood pressure in the microcirculation; the result may be an impedance to blood flow through the wall of the organ. The parasympathetic transmitter, acetylcholine, is a vasodilator substance, but it has not been demonstrated that parasympathetic nerves actually terminate on blood vessel walls.

Stimulation of sympathetic nerves to gastrointestinal organs, as for example during exercise, causes contraction of the smooth muscle of the precapillary segment of the microcirculation with a consequent decrease in blood flow through the organ. This response involves the α-adrenergic receptors in the vascular wall. The response is short-lived and blood flow returns toward normal despite continued sympathetic stimulation for several reasons: 1) β-adrenergic receptors are also activated and they mediate relaxation of the wall of the vessel by stimulating the enzyme adenylate cyclase in the muscle; 2) intestinal precapillary smooth muscle becomes refractory to the constrictor effects by an intrinsic mechanism of the muscle which is called "escape"; 3) reduction in blood flow causes local hypoxia which relaxes constricted vascular smooth muscle; 4) sympathetic stimulation also relaxes the wall of hollow organs, which reduces the resistance to the flow of blood through the wall; and 5) sympathetic stimulation elsewhere in the body causes increases in systemic arterial pressure, cardiac output and the venous return of blood to the heart which tend to increase the flow of blood through gastrointestinal organs.

Circulating Neurohumoral Substances

Several neurohumoral substances are able to influence gastrointestinal blood flow. These include the following agents which reduce blood flow: vasopressin from the posterior pituitary, catecholamines (epinephrine, norepinephrine, dopamine) from the adrenal medulla and angiotensin II from the kidneys. The following circulating hormones have been found to increase gastrointestinal blood flow: glucagon from the pancreas, cholecystokinin from the intestinal mucosa and gastrin from the antral mucosa.

Tissue Metabolism

General hemodynamic factors, the autonomic nerves and circulating vasoactive substances constitute extrinsic influences on gastrointestinal blood flow. There are also intrinsic factors which influence the gastrointestinal circulation, namely local metabolic and vascular phenomena. When there is enhanced metabolic activity by the visceral parenchymal cells (for example, during secretion or absorption) changes occur in the extracellular environment which affect the local smooth muscle cells of the blood vessels. Increased metabolic activity consumes oxygen and generates release of metabolites which relax vascular smooth muscle and increase local blood flow. Examples of these dilator metabolites include carbon dioxide, potassium, ATP and cyclic AMP. Thus, as the cells consume substrate and require more nutrient, they produce materials which will regulate an increase in their own blood flow.

Intrinsic Vascular Phenomena

There are four vascular phenomena of the gastrointestinal circulation which appear to be intrinsic properties of the blood vessels themselves: they are termed "autoregulation," "escape," "redistribution" and the "countercurrent mechanism." Autoregulation is the capacity of a circulation to sustain an even blood flow in the face of widely varying arterial pressures. The smooth muscle of the resistance segment of the microcirculation relaxes to correct a lowered blood flow and lack of oxygen. This characteristic is observed in the circulation of the small intestine and has an obvious protective value, since continued health of an organ depends upon continued blood flow. Thus, a reduction in systemic arterial blood pressure of 10 mm Hg evoked by hemorrhage elsewhere in the body will not significantly affect intestinal blood flow. Autoregulatory protection can be considerable: the vessels supplying blood to the villi can sustain steady flow despite a 70% reduction in arterial pressure.

The escape phenomenon is another compensation to hold blood flow close to normal in the face of the constrictor influence of continuous sympathetic stimulation or prolonged infusion of catecholamines or angiotensin II (Figure 4.3). The response to these constrictors consists of an initial decrease in blood flow to the gut followed by partial restoration of flow to values observed prior to constriction. Escape from sympathetic constriction is the result of several factors. As blood flow to the wall of the gut is reduced, there is an accumulation of dilator metabolites (amines, ATP, etc.) and a reduction in tissue oxygen which relaxes vascular smooth muscle. This effect is greater in the submucosa than in the mucosa and blood flow is distributed away from the mucosa and particularly away from the villi. Metabolites also relax precapillary sphincters and venous smooth muscle further lowering the resistance to blood flow through the

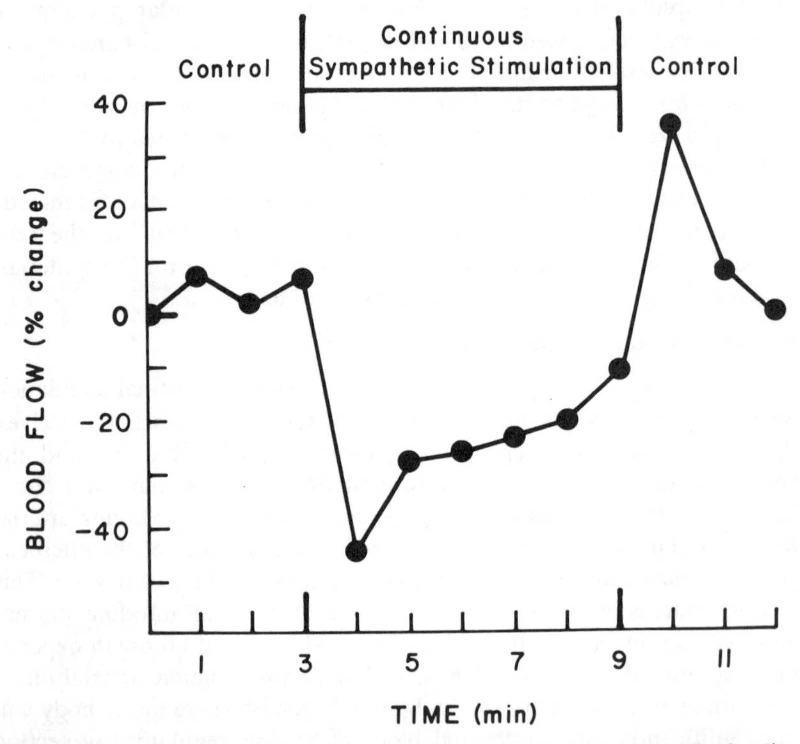

Figure 4.3. The escape phenomenon. Despite continuous sympathetic stimulation the reduction in blood flow through the mesenteric arteries is not fixed at a low level and exhibits a gradual increase toward the flow that prevailed before nerve stimulation.

wall of the intestine, increasing capillary exchange and enlarging the capacitance of the circulation.

Blood flow can be redistributed within the wall of the gut, either from superficial to deeper layers of the mucosa or from the mucosa to the muscularis. Redistribution does not change total blood flow to the organ. Consequently, the angiographic finding of an apparently normal blood flow through the superior mesenteric artery does not invariably mean that the mucosa of the intestine is receiving a normal flow of blood. In fatal, nonocclusive ischemic disease of the gut, the mesenteric artery is patent despite severe ischemia of the mucosa. Sympathetic nervous activity is the major stimulant for redistribution of blood flow away from the mucosa.

Countercurrent vascular mechanisms exist in different parts of the body. The major structural requirements appear to be the close proximity of two blood vessels carrying flow in opposite directions and a difference in the concentration of solutes dissolved in the blood of each vessel. This mechanism will be discussed in more detail in a subsequent portion of this chapter. A point to note is that dissolved oxygen can diffuse from the arterial inflow vessel of the villus to the venule without reaching the capillaries at the tip of the villus. This creates a gradient for oxygen from the base to the tip of the villus and may contribute to the normal demise of surface epithelial cells at the tip where rapid turnover occurs. Since this shunting of oxygen is enhanced by a slowing of blood flow, the countercurrent exchange serves to aggravate tissue hypoxia in ischemic conditions of the bowel.

Based on the preceding considerations about factors which regulate gastrointestinal blood flow one can apply this information to the analysis of a disease involving severe diminution of blood flow in one of the digestive organs. About 3% of all deaths in this country have been ascribed to ischemia of the bowel, which is also termed "hemorrhagic necrosis of the gut."

Intestinal Ischemia

Typically intestinal ischemia occurs in patients with congestive heart failure in which digitalis-type drugs are being used. The patient suffers the sudden onset of severe and constant abdominal pain and is found to be in a state of circulatory shock with abnormally low systemic arterial blood pressure and cardiac output. The diminished blood pressure and cardiac output contribute to a decline in blood flow to the gut. With slowing of flow the viscosity of blood is increased and blood pools, leading to the development of microscopic thrombi in small vessels. Hypotension stimulates release of circulating pressor agents, such as catecholamines, vasopressin and angiotensin II, and there is an increase in sympathetic nervous activity. Heightened tone in vascular smooth muscle combined with a declining arterial pressure initiate collapse of small blood vessels. The net effect of these catastrophic events overwhelms protective phenomena such as autoregulation and escape and evokes a great increase in the resistance to the flow of blood through the intestine. Blood flow to the gut declines further. Sympathetic stimulation redistributes blood flow away from the mucosa and the countercurrent exchanger prevents critical amounts of oxygen from reaching the tips of the villi. Necrosis of the mucosa occurs, starting at the villous tips, and spreads into deeper layers of the mucosa and gut wall. The bowel becomes incapable of preventing its contents from entering the circulation and there is absorption of toxic materials such as bacteria, viruses, exotoxins and endotoxins, lysosomal hydrolases and cathepsins, toxic amines and polypeptides, and substances which depress the function of the heart. The superimposition of a profound toxemia proves fatal.

Until recently this disorder was invariably fatal. It has been possible, in at least a few cases, to insert an intravascular catheter into the superior mesenteric artery and infuse powerful vasodilator drugs to reverse the ischemia and prevent the fatal outcome.

Circulation of Specific Gastrointestinal Organs

SALIVARY GLANDS

The manner in which portions of the gastrointestinal circulation behave relates to the physiology of the particular organ to which the blood vessels are distributed. In the salivary glands the autonomic nervous system is the major regulator of the secretion of saliva. When this secretion is stimulated, glandular blood flow increases, and when secretion abates so does blood flow. Stimulation of α-adrenergic receptors by catecholamines constricts the blood vessels; however, these agents also stimulate the β-adrenergic receptors which are predominant and elicit an increased blood flow to the glands. The parasympathetic transmitter (acetylcholine) is a direct vasodilator substance. During salivary gland secretion vasodilator metabolites are also released locally and further augment blood flow.

STOMACH

In an anesthetized animal at rest blood flow to the mucosa of the stomach exceeds that of the muscularis. The mucosal capillaries appear rather permeable since the concentration of protein in gastric lymph is about half the concentration in arterial plasma. A measurable index of capillary surface area is called the "capillary filtration coefficient." This index has a 10-fold higher value in the gastric mucosa than in skeletal muscle.

Plasma skimming takes place in the gastric mucosa and results in perfusion of the tissue lining the lumen of the stomach with blood having a hematocrit half that of arterial blood. High hematocrit blood flows through vessels located in the deeper layers of the mucosa.

Circulatory functions of the stomach correlate with secretory activity to a considerable degree. When gastric secretion is stimulated by secretagogues such as acetylcholine, gastrin or histamine, there is a corresponding increase in gastric mucosal blood flow (Figure 4.4). Similarly, when gastric secretion is diminished by secretory inhibitors such as atropine, histamine H-2 receptor antagonists or secretin, there is an associated decrease in mucosal blood flow. Agents which alter blood flow to the gastric mucosa operate through at least one of two mechanisms: first, they

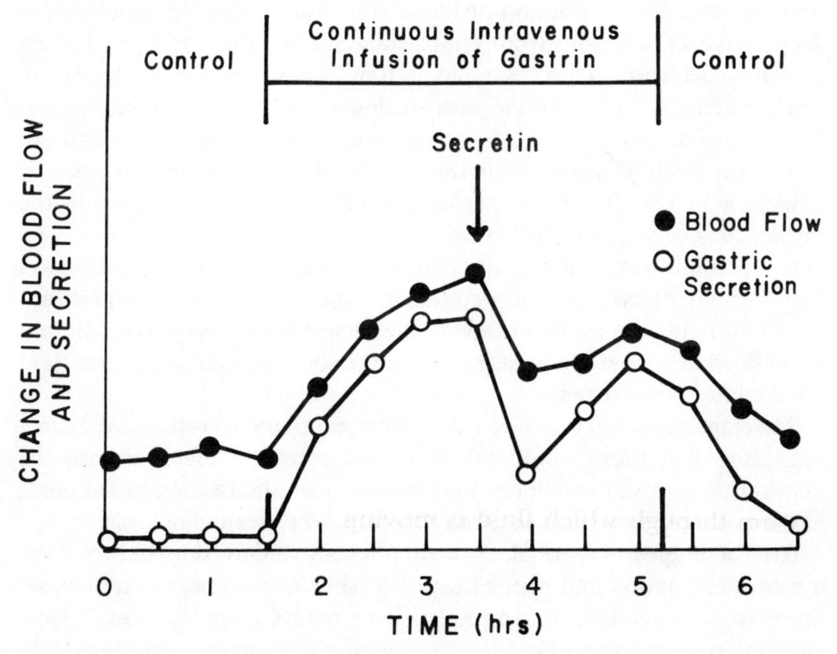

Figure 4.4. The relationship between secretion and blood flow in the stomach. A continuous intravenous infusion of gastrin causes both secretion to commence and blood flow to increase. Partway through the experiment an injection of secretin inhibits secretion and reduces blood flow.

alter secretion and thereby alter tissue metabolism which regulates local release of vasodilator metabolites (carbon dioxide, ATP, etc.); and second, some secretagogues (histamine) are also direct dilators of vascular smooth muscle just as some secretory inhibitors (vasopressin, norepinephrine) are vasoconstrictors. Thus, the two major controlling factors for mucosal blood flow are the rate of secretion of gastric juice accompanied by production of dilator metabolites and the dilator or constrictor characteristics of circulating agents. The parasympathetic nerves and the hormone gastrin operate through the first of these mechanisms, that is, they stimulate increased secretion of acid, pepsin, mucus or gastrin; the by-products of this synthetic machinery are the local dilator metabolites. Vagal stimulation increases gastric mucosal blood flow.

Splanchnic sympathetic nerves are direct vasoconstrictors but also operate through the metabolic mechanism, since adrenergic transmitter substances inhibit gastric secretion and reduce the release of dilator metabolites. Sympathetic stimulation of gastric arteriolar smooth muscle causes constriction and up to a 40% decrease in total blood flow to the stomach. Since redistribution of blood flow away from the mucosa into deeper layers also results from sympathetic stimulation, the overall effect is a 70% decrease in mucosal blood flow. These profound effects are ameliorated somewhat by the escape phenomenon. Sympathetic stimulation also constricts venous smooth muscle, thereby squeezing blood out of the capacitance vessels. Another effect of sympathetic activation is reduction in the density of perfused capillaries and consequently the surface area of the exchange vessels.

It is possible with some dilator drugs to increase gastric mucosal blood flow without affecting gastric secretion. On the other hand, any significant degree of reduction in blood flow to the secreting mucosa will result in a rapid decline in the rate of secretion as the source of nutrients and fluid for the juice is restricted.

Secretagogues evoke an increase of the capillary filtration coefficient indicating that these stimulants of gastric secretion increase both the number of perfused capillaries and the surface area of membrane pores in the capillary endothelium through which fluid is moving.

It is of interest to speculate on a possible relationship between the mucosal circulation and peptic ulcer disease. Ulcers appear in a mucosal lining whose resistance to damage has been reduced. An inadequate local circulation would contribute to the lowering of mucosal resistance. Observations, however, suggest that mucosal circulatory insufficiency occurs subsequent to the critical events that generate peptic ulcer disease. Most topical damaging agents do not have marked vasoactive activities, and ulcers can be produced by either dilator or constrictor drugs. Furthermore, ligation of 90% of the blood supply to the stomach does not induce ulcers. The necessary requirements for an ulcer seem to reside in the corrosive components of gastric juice. In a small fraction of our population the normally resistant lining of the stomach and duodenum fails and ulcers result. The pathophysiological process which leads to peptic ulcers is still unknown.

PANCREAS

The pancreatic microcirculation is arranged so that a portion of its capillaries pass initially to the islets of Langerhans before continuing on to the acini en route to the venules. Other capillaries pass directly to the acini. This is another instance of a part of the gastrointestinal circulation being organized both in series and in parallel with itself. During maximal vasodilation the capillary filtration coefficient is increased 5-fold (along with comparable increases in blood flow) and its rate of filtration per unit weight of organ approximately equals the glomerular filtration rate of the kidney. The high permeability of the pancreatic capillaries may be responsible for the high secretory rates which this small organ maintains. (The pancreas secretes more than 10 times it own weight in pancreatic juice each day.)

The sympathetic nerves are vasomotor to the pancreas and their stimulation reduces blood flow, a response mediated by α-adrenergic receptors. After cessation of nervous stimulation there is a subsequent increase in blood flow (hyperemia) mediated by the β-adrenergic receptors. Vagal nerve stimulation relaxes vascular smooth muscle in the pancreas, but most of the hyperemia is secondary to the effect of the vagus nerves on secretion, metabolism and release of local dilator substances. The hormones, secretin and cholecystokinin, increase the rate of pancreatic exocrine secretion and pancreatic blood flow.

The pancreatic circulation, like that of other gastrointestinal exocrine organs, is linked loosely to secretion. As secretion commences there is an increase in the release of local dilator metabolites and an increase in blood flow to the organ; inhibition of secretion results in a decrease in blood flow. It is not certain that secretory stimulants for the pancreas are released into the circulation in quantities sufficient to act directly on blood vessels independent of the metabolic changes occurring during secretion. Nevertheless, it should be noted that several physiological secretagogues of the pancreas (gastrin, cholecystokinin and acetylcholine) are dilator substances, that is, they relax arterioles and increase mesenteric blood flow. Conversely sympathetic stimulation and circulating catecholamines not only reduce blood flow but also restrict the nutrient and fluid available for the secretory process, thereby reducing secretion.

After the eating of a meal there is an increase in cardiac output and blood flow is augmented to all organs including the pancreas. Eating also activates the vagus nerves, which stimulate secretion of acid by the stomach and the emptying of that acid and ingested food into the duodenum. Acid and food evoke the release of secretin and cholecystokinin from the mucosal lining of the duodenum. Release of the parasympathetic transmitter, acetylcholine, and the two peptide hormones stimulates pancreatic secretion and metabolism, causing release of local dilators. The result of eating is increased pancreatic blood flow.

LIVER

The circulation of the liver receives contributions from both arterial and venous sources. About three-quarters of the blood supply of the sinusoids comes from the portal vein with the remainder is derived from the hepatic artery. As blood enters the hepatic artery its pressure is about 100 mm Hg and is reduced to less than 10 mm Hg by the precapillary resistance vessels of the circulation. Normally, portal vein pressure is slightly greater than 10 mm Hg. In cirrhosis of the liver, fibrosis and shrinkage of the organ impede venous flow and elevate portal pressure. In severe cases there will be engorgement of the branches of the portal vein, such as those draining the lower esophagus. These esophageal varices are vulnerable to erosion by gastric juice and may be the source of torrential hemorrhage leading to death.

The sinusoids are the vessels in which the arterial and venous blood mix. The endothelial lining of the sinusoids is more permeable to protein than is the endothelium of capillaries elsewhere in the body. The sinusoids are also the site where exchange of substrates and metabolites occurs between adjacent parenchymal cells.

Sympathetic stimulation of the hepatic circulation increases precapillary segment resistance on the arterial side and lowers blood flow through the hepatic artery. Sympathetic stimulation also constricts the smooth muscle of the portal vein raising pressure in that vessel and forcing blood to flow through the liver at a faster rate. Under resting conditions the liver is a repository of a large volume of noncirculating blood being held in venules and other capacitance vessels. Sympathetic stimulation squeezes this blood out of the organ and returns it to the heart. In congestive cardiac failure there is an elevated pressure in the systemic veins which is transmitted backwards into the capacitance vessels of the liver. The liver becomes engorged with massive volumes of noncirculating blood, and the organ is palpably enlarged as a result (hepatomegaly).

SMALL INTESTINE

General Considerations

In the resting small intestine 75% of total blood flow is distributed to the mucosa with the remainder passing to the submucosa and the muscularis. Since the muscular layer comprises one-half the total mass of the wall its blood flow per unit weight is lower than flow to the other two layers. Vasodilator drugs increase intestinal blood flow greatly, especially perfusion of the mucosa; thus, with some drugs flow to the villi alone could comprise over half of the total intestinal blood flow, leaving less than one-tenth of the total blood flow to perfuse the muscularis.

At rest the capillary filtration coefficient of the gut is about 10 times higher than that of skeletal muscle reflecting both a greater density of perfused capillaries and a greater capillary pore area for filtration. The intestinal capillaries exhibit a high permeability to most solutes. Although not viewed principally as a secretory organ, in matter of fact, the small intestine is capable under a number of circumstances of secreting large volumes of fluid. The source of this fluid is the exchange segment of the circulation, namely the capillaries with their vast surface area. At rest only one-third of the capillaries of the villi are open to the flow of blood. Certain drugs can relax the precapillary sphincters and open nearly all capillaries for perfusion, resulting in a greatly increased exchange of materials between blood and cells. Under such conditions more oxygen is available and consumption of oxygen increases.

The volume of blood actually held in the mucosa can be more than doubled either by administering dilator drugs, by eating a meal or by increasing venous outflow pressure from 10 to 25 mm Hg. Over half of the total small intestinal blood volume is contained in the venules and small veins of the gut.

Countercurrent Mechanism

In the villi a countercurrent exchange mechanism appears to involve the microcirculation. This is depicted ideally in Figure 4.5. As indicated previously the anatomical arrangement of blood vessels in the villus seems to permit the loss of physically dissolved oxygen from the arterial inflow vessel to the outflow vessels. In addition, a recent concept about the countercurrent arrangement in the villus is that it serves as a multiplier for the generation and preservation of a standing concentration gradient

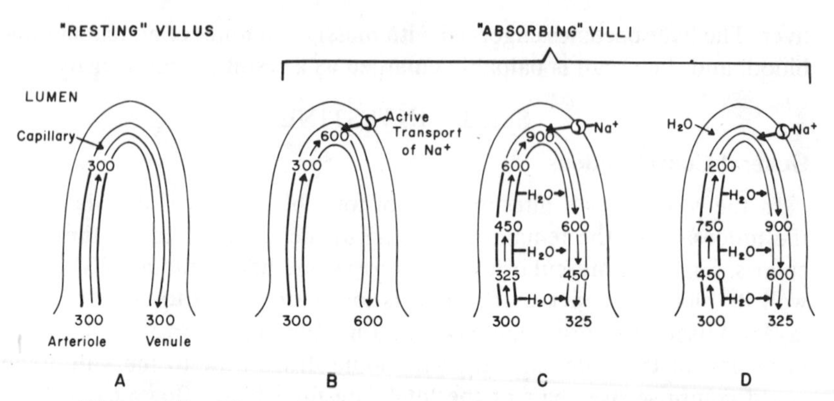

Figure 4.5. The countercurrent multiplier in the intestinal villus.

for solutes along the length of the villus. A result of such a mechanism would be an interstitial osmolality of up to 1200 mOsmol/kg at the tip of the villus analogous to what is found in the papilla of the kidney. The minimal requirements for countercurrent multiplier are the close proximity of two arms of a hairpin loop each conveying fluid in opposite directions and the presence of an active transport mechanism for sodium. The inflowing arteriole is one arm of the hairpin loop. At the tip of the villus, the arteriole arborizes into a network of capillaries which carry the blood in the opposite direction toward the base of the villus. The epithelial cells covering the surface of the villus possess a well-developed transport system for sodium which is capable of moving that ion against electrochemical gradients and is energy dependent.

A schematic sequence of events which generate the countercurrent multiplier in the villus appears in Figure 4.5.

In theory the "resting" villus would show a concentration of osmolytes in the blood flowing in and out which was uniform at 300 mOsmol (Figure 4.5*A*). The onset of the active transport of sodium by the epithelial lining of the villus adds that ion to the blood in the small vessels inside the villus, thereby increasing the concentration of osmolytes to 600 (Figure 4.5*B*). As the higher concentration drains down the capillary an osmotic gradient is produced between the inflowing and the outflowing vessels of this hairpin loop. These changes in solute concentrations attract water from the inflowing arm to the outflowing arm of the hairpin loop (Figure 4.5*C*). Consequently, the concentration of osmolytes increases as blood flows toward the hairpin and decreases as blood flows away from the hairpin loop. Meantime, the sodium pump is adding that ion to the circulation, which raises concentrations still further. In the steady state the concentration of osmolytes at the outset is 300, even as it was in Figure 4.5*A*. As the blood flows down the arteriole toward the hairpin, water is osmotically removed from the inflowing blood, thereby raising the concentration up to 1200 as shown in Figure 4.5*D*. As the blood flows through the capillary on its way from the hairpin the concentration gradually falls to 325 as the blood leaves the villus. The reason that the concentration does not exceed 1200 at the tip of the hairpin loop is that water as well as sodium is being attracted into the villus from the lumen of the gut.

Since there is an interstitial osmolality at the tip of the villus 4 times greater than at the base, the lining of the gut offers a powerful osmotic attraction to water in the lumen. This standing interstitial gradient is dissipated when there is a great increase in blood flow, since the transit time of blood would be reduced 80% in the villus. The advantage of a compartment in the villus which is hyperosmolar lies in the reduction of the work involved in absorption of water. The expenditure of energy in transporting a much smaller number of sodium particles to generate the osmotic gradient results in the movement of a larger number of water particles during usual absorptive situations. The low cost of work in generating and maintaining an interstitial osmolality of 1200 mOsmol/kg in the distal villus depends partly upon the structural arrangement of the microcirculation of the villus, namely a hairpin loop. The result of these physiological and anatomical features is a remarkable absorptive feat—each day the small intestine absorbs more than 8 liters of water with the expenditure of minimal amounts of energy.

Other Influences on Blood Flow

The sympathetic nerves are the only vasomotor fibers distributed to the intestine. Stimulation of these nerves produces constriction of smooth muscle in the resistance and capacitance segments as well as in the precapillary sphincters with a rapid reduction in total blood flow to the small intestine. The decline is transitory, however, and intestinal blood flow is partially restored to control levels within 2 to 4 minutes after the onset of continuous nerve stimulation. The escape from sympathetic constriction occurs in each segment of the intestinal microcirculation; consequently, the reduction in intestinal blood flow, capillary filtration coefficient and blood volume are only temporary responses to sympathetic stimulation.

At rest blood flow to the small intestine constitutes about half of the splanchnic flow and this proportion is increased during eating as intestinal blood flow is doubled. Cardiac output also increases with eating for a short period and food in the gastrointestinal tract evokes increases in blood flow and prolongs the duration of dilation for 3 or more hours. The bulk of these effects appear to be mediated by the central nervous system through the vagus nerves and by gastrointestinal hormones which stimulate metabolism of various tissues during feeding and initiate the release of dilator metabolites. There may also be some direct dilator effects of the gastrointestinal hormones themselves. Part of the dilator effect is triggered by the hyperosmolarity of the food and there also appears to be some release of specific local dilator substances (serotonin and histamine) in the mucosa when food is in the lumen of the gut.

Besides feeding, other normal activities also alter intestinal blood flow. Exercise reduces blood flow to the gut via activation of the sympathetic nervous system. The spontaneous mechanical movements of the wall of the gut (motility) change wall tension and can generate sizable pressure increases in the wall exceeding the pressure in the microcirculation. This uncommon occurrence will impede blood flow transiently.

The major function of the small intestine is absorption of digested materials from the lumen of the gut. Absorption rates for passively transported solutes which are lipid-soluble vary directly with blood flow, whereas absorption of water-soluble compounds such as urea is relatively independent of blood flow. These findings are predictable based on the known permeability characteristics of the epithelial lining of the gut to the two different kinds of passively transported materials: lipid-soluble substances readily cross the lining of the gut, whereas water-soluble compounds of molecular weight greater than about 100 daltons permeate poorly. Therefore, for water-soluble substances the passage across the epithelium (rather than the blood flow) is the rate-limiting absorptive step, whereas blood flow is a rate-limiting step for lipid-soluble substances. Actively transported substances also depend on blood flow, since blood is not only the transport vehicle for the solute but also is the source of nutrients for metabolism of the absorptive cells.

Absorption of gases by the mucosa of the small intestine varies according to the site in the mucosa where the gas is being absorbed. Blood vessels close to the lumen determine the disappearance of gases, such as hydrogen. Here absorption of the gas is flow-limited. Blood vessels further away from the lumen in the mucosa do not influence the absorption of a gas to the same extent, that is, absorption of gases through these vessels is limited by the rate of diffusion of the gas from the lumen to the more distant blood vessels.

Shock

An overwhelming disease state which affects the intestinal blood vessels is circulatory shock. In the severe forms of this disorder the circulation of the gut becomes a prime target, and the breakdown of its functions leads to necrosis of the intestine and to death of the patient. As the postcapillary segment deteriorates, the venules and capillaries begin to sequester large volumes of blood which stagnate the lost fluid to the interstitium, further aggravating the hypovolemic situation in shock.

COLON

Resting colonic blood flow is somewhat lower than that of the small intestine, but the filtration coefficient and the blood volume are of the same magnitude. A major portion of blood flow is distributed to the mucosa. Parasympathetic pelvic nerves dilate the mucosal blood vessels. The vascular mediator released by these nerves is bradykinin, a potent vasodilator. Sympathetic stimulation evokes a transient reduction in blood flow which is followed by escape.

Summary

The anatomical organization of the gastrointestinal circulation is complex with frequent examples of vessels being arranged both in series and in parallel. These arrangements permit such functional features to emerge as redistribution of blood flow between layers of the wall of hollow organs. The main forces influencing blood flow to gastrointestinal organs include general circulatory factors in the body, autonomic nerves, circulating vasoactive materials, products of tissue metabolism and unique features of the microcirculation such as autoregulation, escape, redistribution of blood flow and the countercurrent mechanism.

Blood flow to the salivary glands is regulated primarily by acetylcholine and catecholamines. Blood flow in the stomach is related to the major secretory function of that organ. Thus, when secretion of acid gastric juice is increased, there is an increase in gastric mucosal blood flow. Similar relations between blood flow and secretion appear to exist in the pancreas. The major influence on blood flow to the liver appears to be the sympathetic nervous system. The sympathetic nerves also reduce blood flow to the small intestine, but the effect is transient. There is a countercurrent multiplier in the intestinal villi which is partly responsible for establishing an hyperosmolar interstitium at the tip of the villi. This standing osmotic gradient down the length of the villi results in rapid absorption of water from the lumen of the gut.

chapter 5

Gastrointestinal Hormones

how these blood-borne substances originate, their chemical composition and their effects on gastrointestinal organs

A hormone is a chemical which is secreted into the blood in response to a physiological stimulus and travels to a site distant from the place of its release to cause a physiological response. The gastrointestinal tract is the locus for production and release of many hormones, including some which act uniquely on gastrointestinal organs, some which act on all organs of the body and some whose actions have yet to be determined. Four gastrointestinal hormones are known to play an important role in the regulation of gastrointestinal organs specifically, namely gastrin, cholecystokinin (CCK), secretin and gastric inhibitory peptide (GIP). When food is ingested these four hormones are released in an orderly manner as the food is propelled from the stomach into the intestine, and their release activates or inhibits secretory and motor activity of various organs. The end results are well-controlled processes or propulsion and digestion of food. The events involved in this interaction between hormonal release and organ function in the upper gastrointestinal tract are depicted in Fig. 5.1.

The gastrointestinal hormones share some common features despite differences in chemical structure, cell of origin and biological actions. The common characteristics include their polypeptide composition, their origin in specialized granular cells of the gastrointestinal epithelium and their

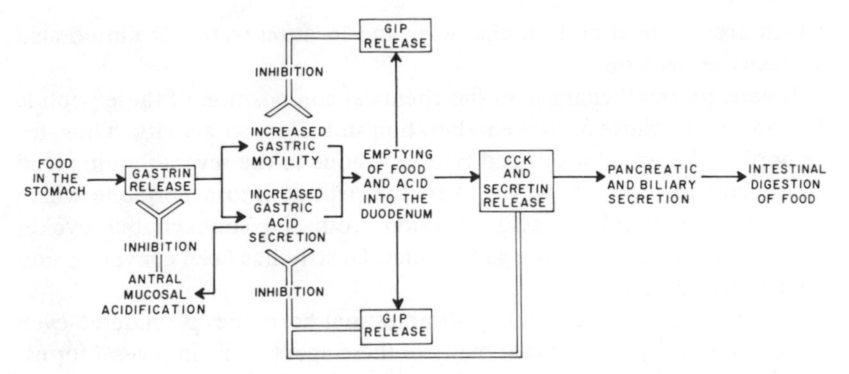

Figure 5.1. Effect of food on the release of gastrointestinal hormones.

ability to influence secretory and motor functions of several gastrointestinal organs in selective ways. For the hormones gastrin, secretin and CCK, there is also the phenomenon of opposing effects on certain organs, *e.g.,* the gastric secretory response to gastrin is inhibited by secretin and by CCK. Another kind of interaction between these hormones is "synergistic," *e.g.,* the stimulatory action of secretin on pancreatic secretion of fluid is enhanced by CCK.

Although all of these hormones evoke significant responses from the organs of the gastrointestinal system when administered in large doses, only one of the agents has been truly implicated in a disease caused by excessive production and release of that hormone. This disorder is called the Zollinger-Ellison syndrome and the culprit is the hormone gastrin. However, since the origin of many gastrointestinal disorders is unknown, it is not unlikely that in the future other diseases will be demonstrated with a hormonal etiology.

The gastrointestinal hormones are members of two families of compounds. The first family includes gastrin and CCK. Despite the dissimilar size of these two polypeptide hormones, the C-terminal five amino acids of each are identical in both nature and sequence. Furthermore, all of the biological activities of gastrin and CCK are present in the C-terminal tetrapeptide, that is, the two hormones share similar actions but exhibit different degrees of effectiveness in their actions on different organs of the gastrointestinal tract. The second family of gastrointestinal hormones consists of secretin, GIP, vasoactive intestinal peptide and glucagon. The chemical similarities of these hormones are not quite as striking as in the case of the first family. Thus, glucagon has 29 amino acids, nearly half of which are identical both in character and location to the 27 amino acid sequence of secretin.

Relatively small changes in the chemical composition of these peptide hormones can cause a marked alteration in biological activity. Thus, for example, addition of a sulfated tyrosyl residue to the seventh amino acid of the gastrin molecule will convert the peptide into a hormone which only weakly stimulates acid secretion from the stomach but evokes powerful contractions in the gallbladder. Gastrin has been converted into a CCK-like hormone.

The chemical nature of the gastrointestinal hormones is rendered even more complex by the fact that many of these agents exist in several forms. For example, the hormone gastrin exists in three distinct forms, namely a 17-amino acid form called "little gastrin" or G17 gastrin, a 34-amino acid form called "big gastrin" or G34 gastrin and a molecule referred to as "big big gastrin." These forms of the hormone have different potencies and are present in different amounts in the body. Their distinction has been made possible by radioimmunoassay.

When a hormone has been extracted from the gastrointestinal mucosa, purified and even synthesized, it becomes possible to harvest specific antibodies in animals which are used either to measure the concentration of hormone that is present in serum or to localize the cells in the body which produce the hormone. In the latter case the antibody is coupled with a fluorescent dye or other label which can be discriminated with a light or electron microscope. There are certain cells in the mucosae of the stomach and gut which bind particular antibodies and dye. These cells have granules in them which are characteristic of endocrine cells elsewhere in the body. These cells are involved in amine precursor uptake and decarboxylation (APUD series) and were derived from the neural crest of the embryo.

Radioimmunoassay of the hormone gastrin is commonplace in American hospitals because the measurement may provide definite proof of the presence of the Zollinger-Ellison syndrome. The assay involves quantitative binding of antigen (hormone) and its antibodies. It is an extremely sensitive method which can detect picomoles of the hormone.

Gastrin

The most important form of gastrin is the G17 molecule. The major source of the hormone is the G cell of the antral mucosa. All of the actions of gastrin are present in the C-terminal tetrapeptide moiety of the molecule. There is a commercially synthesized chemical consisting of the tetrapeptide to which has been added a stabilizing amino acid; it is called "pentagastrin" and has been used extensively in human subjects.

The gastrointestinal tract responds in many ways to the administration of large doses of gastrin. This hormone causes contraction of the lower esophageal sphincter, enhanced gastric secretion of acid and increased motor activity of the stomach. Gastrin is also a potent stimulus for exocrine secretion of enzymes by the pancreas, but only a weak stimulus for the secretion of bicarbonate and water by that organ. In addition, gastrin will evoke release of insulin from the islet cells of the pancreas. Gastrin increases biliary secretion of bicarbonate and duodenal secretion of water and mucus from Brunner's glands. The hormone slows the absorption of sodium, potassium, glucose and water from the gut. It increases motility of the intestine and gallbladder. Gastrin is a "trophic" substance for the mucosal lining of the stomach, small intestine, colon and pancreas, *i.e.*, the hormone increases DNA, RNA and protein synthesis of these tissues. Gastrin also stimulates contraction of the uterus and is a vasodilator agent in the mesenteric and gastric circulations.

These numerous responses to a large amount of gastrin might be termed the "pharmacological" actions of the hormone. When gastrin is administered in doses which produce concentrations in the blood equivalent to those normally found, however, only a few of the preceding actions were observed. These might be termed the "physiological" responses to the hormone. Physiological responses to the hormone are diminished considerably when gastrin is removed from the body, either by surgical removal of the antral portion of the stomach or by depriving a patient or animal of food (but maintaining nutrition by means of parenteral alimentation). Under circumstances of gastrin deficit the primary effects are a great reduction in the ability of the stomach to secrete acid and an atrophy of the gastrointestinal mucosa. It appears that the main physiological actions of gastrin are stimulation of acid secretion by the stomach and its trophic effect, *i.e.*, acting as the specific growth hormone for the mucosa of the stomach, small intestine and colon.

The major source of gastrin is the antral mucosa. When the pH of the fluid bathing the antral mucosal surface rises above 2.0, gastrin is released from the G cells into the blood draining the region and is carried to the liver and the heart to be pumped to all organs of the body. Gastrin reaches the stomach, where it stimulates secretion of acid, and the gastrointestinal mucosa, where it maintains the integrity of the tissue. When the pH of the fluid bathing the antral mucosa falls below 2.0, gastrin is no longer released. Food has a pH that is above 2.0 and ingestion of a meal will usually cause an increase in the release of gastrin. When food has been emptied from the stomach and acid is being secreted briskly, there is a decline in pH at the antral surface providing an effective feedback to turn off gastrin secretion. This negative feedback control of gastrin release by antral acidity is shown in Fig. 5.1.

Hypergastrinemia is observed in the Zollinger-Ellison syndrome. The source of excess gastrin in this disease is an APUD cell in the islets of Langerhans of the pancreas. There is overgrowth of an endocrine cell in the islets which produces gastrin. The overgrowth of these cells occurs either as a discrete tumor of the organ ("apudoma") or as diffuse hyperplasia. Beside the continually elevated levels of the hormone in the blood, there are some life-threatening consequences. These include secretion of massive volumes of highly acidic gastric juice each day. Normally an adult male will secrete about 1 to 2 liters of juice daily containing 100 mEq or less of hydrogen ion; in the Zollinger-Ellison syndrome these amounts of fluid and acid may be increased 5-fold. Chronic bathing of the duodenal mucosa with such large volumes of acid cause ulcers which will not heal and will often perforate. Exposure of the gut to large volumes of fluid produces unremitting diarrhea with loss of potassium that may lead to hypokalemia and renal tubular damage. Until recently the only effective treatment for this disease has been surgical removal of the acid producing portion of the stomach. Now the disease may be managed with a drug, *i.e.,* through administration of a histamine H-2 antagonist (cimetidine).

Secretin

When acidic gastric juice is emptied by the stomach and contacts the mucosal lining of the duodenum, a secretory response of the pancreas ensues in which copious volumes of fluid laden with bicarbonate are produced. Denervation of either the intestine or the pancreas does not prevent this response. The blood-borne mediator of the pancreatic secretory response to acid in the intestine is the hormone secretin, which, incidentally, was the first hormone discovered.

Secretin is a polypeptide containing 27 residues. Removal or exchange of any of these 27 amino acids causes loss of the biological activity of the hormone, *i.e.,* secretin has no "active site" as does gastrin. Secretin is produced by specialized APUD cells localized to the mucosa of the duodenum and jejunum.

A decline in the pH value of the fluid bathing the upper intestinal mucosa below 4.5 prompts secretin release (Figure 5.1). Release of the hormone does not appear to result from contact by the mucosal lining with protein or carbohydrate.

When secretin is administered in large doses, there are several responses by gastrointestinal organs. These include stimulation of the production of pepsin by the stomach with simultaneous inhibition of the secretion of acid in response to gastrin. This latter action is one of the several situations in which the actions of secretin and gastrin are opposed. Secretin evokes secretion of the inorganic components of bile by the liver. Secretin inhibits gastrointestinal motility, again as opposed to actions of gastrin. Release of insulin is prompted by secretin. The production of cyclic AMP is enhanced in pancreatic acinar cells by secretin, suggesting that this nucleotide is the intracellular mediator of hormonal action. Secretin also opposes the trophic action of gastrin on the gastrointestinal mucosa.

The only one of these preceding actions of secretin which appears to be a response occurring under usual circumstances is the pancreatic secretion of alkaline juice. Presumably the major physiological contribution of secretin, therefore, would be to assure the presence of a pH optimum in the lumen of the upper gut for the powerful digestive enzymes of the pancreas which would be inactivated by gastric juice.

There is evidence that cigarette smoking is accompanied by a significantly elevated incidence of duodenal ulcers and the time of healing of these ulcers is delayed in smokers. Nicotine in cigarette smoke inhibits pancreatic secretion, partly by inhibiting the release of secretin. It is possible that duodenal ulcers may be linked generally to a deficient release of secretin.

Cholecystokinin (CCK)

This hormone contains 33 amino acids, although the entire biological activity of CCK is contained in the C-terminal octapeptide moiety. The body also produces a somewhat larger form of CCK.

Another APUD cell type located in the jejunum particularly, but also in the rest of the small intestinal mucosa, produces granules containing CCK at the basement membrane of the cell. These CCK-producing cells also have processes which extend into the intestinal lumen. The major stimulus for production of CCK is the presence of fat in the lumen of the gut.

Administration of large doses of CCK stimulates contraction of the gallbladder, pancreatic secretion of enzymes, gastric secretion of pepsin, Brunner's gland secretion, biliary secretion of inorganic juice, motility of the stomach and gut and growth of the pancreas. CCK also produces satiety to food. This hormone interacts with others mentioned previously. Thus, CCK is a weak stimulant of gastric acid secretion but competitively inhibits secretion of acid in response to gastrin (Figure 5.1). CCK stimulates release of insulin, dilates the mesenteric artery and enhances the actions of secretin on pancreatic exocrine secretion.

Of the preceding actions of CCK, the two which are best established as physiological are contraction of the gallbladder and stimulation of pancreatic secretion of enzymes. Other likely physiological actions of CCK may turn out to be the interaction with secretin and the growth-promoting action on the pancreas.

There are no known diseases in which either excessive or deficient degrees of production of CCK have been implicated. It has been suggested, however, that morbid obesity may involve deficient production of or a poor hypothalamic response to CCK.

Gastric Inhibitory Peptide (GIP)

GIP is chemically related to secretin and glucagon. The major source of this hormone is the mucosa of the upper small intestine. Its primary action is to inhibit gastric secretion and motor activity. GIP inhibits both acid and pepsin secretin. Gastric emptying is slowed by this hormone. GIP stimulates secretion by the gut and release of insulin from the pancreas. The major component of a meal which stimulates release of GIP is fat, but carbohydrate is also potent in this regard.

GIP is sufficiently insulinotropic that this may be one of the physiological actions of GIP. GIP could, therefore, be the major messenger stimulating release of insulin after a meal enters the gut and some forms of diabetes mellitus may involve GIP. The other major physiological role of GIP is its negative feedback inhibition of gastric secretion and emptying which takes place when the components of a meal have been dumped into the gut by the stomach (Figure 5.1). This inhibition is necessary to brake the rate of gastric activity relative to the capacity of the upper gut to handle fluid and nutrients, as well as leading to the eventual restoration of the stomach to its quiescent interprandial state (between meals).

Other Hormones

Insulin and glucagon originate in gastrointestinal organs. However, their major actions are on all organs of the body rather than serving unique needs of the digestive system. For these reasons their endocrinology will not be discussed in this chapter.

In addition there are a sizable numbers of discrete polypeptides which have been extracted from the mucosal lining of the gut. Some of these have been synthesized and utilized to develop specific radioimmunoassays for their detection in blood. Without exception each of these agents produces some response of the gastrointestinal organs when the agent is injected into the body. What is not known is whether any of these responses is physiological, namely responses occurring under usual conditions specifically as a result of the release of the agent. Therefore, these agents cannot be termed true hormones, but they are candidates for that role. Some of the more promising candidates include vasoactive intestinal peptide (VIP) and motilin.

VIP is found in both the mucosa of the gut and in the brain. It stimulates release of insulin and inhibits gastric acid secretion, much like GIP to which it is chemically related. It also stimulates glycogenolysis, pancreatic secretion of water and bicarbonate, and small intestinal secretions. VIP exhibits an inotropic effect on the heart and is a potent vasodilator agent. Although the physiological importance of VIP has not been settled there is one pathological condition involving this hormone. Tumors of the pancreas and sympathetic ganglia may produce excessive amounts of VIP. The result is severe and unremitting diarrhea, loss of potassium with hypokalemia and achlorhydria, reflecting some of the major actions of VIP. This syndrome is sometimes referred to as "pancreatic cholera."

Motilin is also derived from the upper intestinal mucosa. This hormone has a powerful inhibitory effect on gastric emptying, stimulates secretion of pepsin and opposes growth of the gastric mucosa.

The gastrointestinal tract is the target of hormones which originate primarily elsewhere in the body. Somatostatin is generated primarily in the hypothalamus, elsewhere in the brain and in the gastrointestinal mucosa and pancreas. This hormone has important interactions with pituitary hormones and generates significant gastrointestinal responses, such as inhibition of the release of gastrin, of gastric acid secretion, of pepsin secretion and delay of gastric emptying. Somatostatin also blocks the effects of CCK on the gallbladder and pancreas.

The growth of the gastrointestinal lining appears to depend on the release of the hormone gastrin, whose major site of production is the antral mucosa of the stomach, into both blood and lumen. The antral mucosa, however, does not respond to the growth-promoting effects of gastrin itself. The trophic hormone for the antral mucosa appears to be the somatotrophic hormone of the pituitary. Adrenal cortical hormones of the mineralocorticoid type appear to have effects on intestinal transport of electrolytes similar to effects seen in the kidneys.

Although histamine and prostaglandins are not classified as hormones, they exhibit powerful and specific effects on gastrointestinal organs and both substances are found in high concentration naturally in the mucosa. Histamine stimulates copious secretion of acid by the stomach. A specific receptor on the oxyntic cell, termed the histamine "H-2" receptor, is sensitive to histamine and can be blocked by antagonist drugs like cimetidine which inhibit secretory activity of the stomach. Histamine also increases motor activity in the gut and is a powerful vasodilator substance in the arteries of the stomach and gut.

Prostglandins are potent inhibitors of gastric secretion of acid and pepsin. These agents increase intestinal motor activity, are vasodilator drugs in the gastrointestinal circulation, relax the lower esophagel sphincter and stimulate secretion by the gut. Prostaglandins also confer protection on the gastrointestinal mucosae against various ulcerogenic stimuli. Drugs which deplete mucosal tissue stores of prostaglandin cause ulcers.

Summary

The gastrointestinal hormones originate in specialized endocrine cells of the mucosal lining of the stomach and gut. These hormones are polypeptides in chemical composition. They are released into the blood and circulate throughout the body to return to the gastrointestinal organs where they exhibit their effects. These hormones are powerful stimulants or inhibitors of various gastrointestinal organs when the hormone is injected into the blood in large amounts, far exceeding concentrations seen in the blood under normal circumstances. The physiological actions of these hormones occur when the concentration of hormone in the blood is at the level observed during normal activity.

The four hormones which appear to be physiological regulators of the gastrointestinal organs are gastrin, secretin, CCK and GIP. Gastrin is a stimulant of gastric acid secretion and is a trophic hormone for the gastrointestinal mucosa. Secretin is a powerful stimulant of pancreatic exocrine secretion of water and bicarbonate. CCK is a stimulant of pancreatic exocrine secretion of digestive enzymes. CCK also prompts powerful contractions of the gallbladder. GIP is insulinotropic and is a potent inhibitor of both gastric secretion and emptying.

Chapter 6

Neural Regulation

how the autonomic nervous system influences secretory and motor activity of gastrointestinal organs

The gastrointestinal organs respond normally to nervous stimuli. It is well known that emotional stresses can cause common functional symptoms referable to these organs—symptoms such as belching, heartburn, abdominal discomfort, vomiting, flatulence and bowel movements. There is heavy traffic of impulses traveling to and from the gastrointestinal tract in nerves which are located outside the organs, the so-called "extrinsic" nerves. Afferent messages evoke central nervous responses to events in the digestive organs. Efferent stimuli initiate or influence three kinds of gastrointestinal functions: 1) exocrine secretion in salivary glands, stomach and pancreas; 2) motor effects in each of the hollow organs; and 3) endocrine secretion of gastrointestinal hormones. In addition to the extrinsic nerves there is an extensive network of "intrinsic" nerves in the gastrointestinal organs which connect to the extrinsic nerves and which are also capable of influencing secretion and motility. This complex neural organization is diagrammed in Figure 6.1.

Afferent Nerves

The majority of nerve fibers in the vagal trunks convey impulses from the gut to the brain. Receptors in the walls of hollow organs are sensitive both to stretch and to the chemical and osmotic composition of fluid in

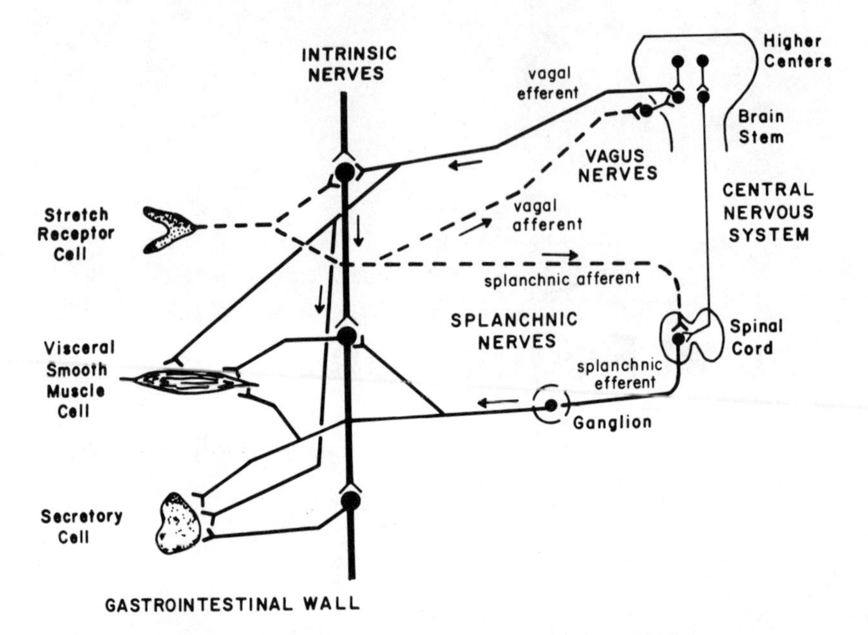

Figure 6.1. Organization of intrinsic and extrinsic afferent and efferent neural pathways to responsive gastrointestinal cells. *Dashed lines* refer to afferent and *solid lines* to efferent nerves or interneurons.

the lumen of the organ. Mild degrees of stretch in hollow organs are usually unnoticed but greater stretch (as by gas) may be percieved as uncomfortable or painful. There are also pain fibers in the substance of the pancreas, liver capsule and mesenteries. Such pain may be sensed in the abdomen or referred to the lumbar region. Afferents conveying painful sensations pass up the vagi or go to dorsal roots in the cord.

Other afferents from the stomach transmit the feeling of hunger pangs. There are also baroreceptors in the arteries of the pancreas which sense hemodynamic changes and relay information through the sympathetic nerves. In addition, afferent messages in one part of an organ may travel to another part through the intrinsic nerves; an example of such an event is the relaxation of the lower esophageal sphincter following distention of the upper esophagus. A longer neural connection involving two organs and the nerves extrinsic to them is found in the increased colonic motor activity which follows ingestion of a meal into the stomach, the so-called "gastrocolic reflex."

Neural Control of Secretion

Many of the gastrointestinal organs contain glandular tissue which secretes a juice that is involved in the digestion of food. Nerves pass to glandular cells or synapse with intrinsic nerves in the tissue and influence the rate of secretion of these juices. The three organs in which this control is best understood are the salivary glands, the stomach and the pancreas. Salivary glands receive innervation from both the parasympathetic and the sympathetic portions of the autonomic nervous system. Release of either acetylcholine (cholinergic nerves) or the catecholamines (adrenergic nerves) causes stimulation of the secretory process in salivary glands, as well as increasing metabolism, blood flow and muscular activity in the salivary ducts.

The stomach receives both parasympathetic and sympathetic innervation. Impulses delivered by the vagus nerves prompt secretion of acid from the oxyntic or parietal cells, pepsinogen from the peptic cells and mucus from the mucous cells of the gastric mucosa. These three secretions constitute important solutes of the gastric juice. In addition, vagal stimulation in the dog causes the release of the hormone gastrin from certain endocrine cells (G cells) located in the antral mucosa of the stomach. This hormone is released into the blood and acts at a distance from the cells of origin. The vagal nerves either terminate directly on the target cells or synapse with the intrinsic nerves of the gastric mucosa. These intrinsic nerves, in turn, release acetylcholine in the vicinity of the oxyntic, peptic, mucous and G cells, thereby stimulating secretion. Certain stimuli in the stomach itself evoke a message which travels up the afferent nerves of the vagi to the brain stem where other neurons connect to the vagal nuclei; the neurons include connections from the hypothalamus, globus pallidus and the limbic system. Glucocytopenia (low glucose content inside cells) is a stimulus for gastric acid secretion and the stimulation is mediated through the aforementioned central neural pathways to the vagal nuclei.

The vagus nerves also terminate in the substance of the pancreas where they synapse with intrinisic cholinergic nerves which then pass to the acinar cells and islet cells of the pancreas. Vagal efferent fibers also terminate directly on these same cells. Release of acetylcholine causes acinar secretion of digestive enzymes into the pancreatic juice as well as release of the hormone insulin from the β cells of the islets. When the pancreas is stimulated to secrete its juice, there is an increase in metabolism of the organ and an increase in blood flow to the organ to supply the needed extra nutrients for the secretory process. Sympathetic nervous stimulation causes a decrease in blood flow to the pancreas and will, therefore, inhibit secretion in the gland.

Neural Control of Motility

Motor activity of the hollow organs of the digestive tract is regulated both by intrinsic nerves located within the wall of the organ and by extrinsic nerves emanating from the central nervous system. There is considerable coordination of these two components of the nervous system in influencing the onset and relaxation of muscular activity as well as the intensity and frequency of contractions. Impulses reach the viscera through both the parasympathetic and sympathetic portions of the autonomic nervous system. These extrinsic nerve fibers terminate either near visceral smooth muscle cells or synapse with intrinsic nerves which themselves also terminate near muscle units. The intrinsic nerves are arranged in interconnecting networks located between layers of smooth muscle. These networks or plexuses contain motor neurons, interneurons and sensory neurons. The plexuses are capable of mediating coordinated propulsive movements of muscle within a short segment of gut; such movement is called "peristalsis" and can occur in the absence of extrinsic nervous input. Thus, a surgically denervated segment of intestine can be stimulated to exhibit peristalsis reflexly by distending the gut from within.

Characteristically, there is considerable spontaneous contractile activity of visceral smooth muscle which is influenced by intrinsic nerves acting to modify the relationship between the basic electrical rhythm (slow waves) and standard action potentials of the smooth muscle.

Both the extrinsic and intrinsic nerves cause both stimulation and inhibition of visceral smooth muscle. Inhibition of motility may occur as the result of:

1. Messages originating in the brain and conveyed to the gut through either the vagi or through the spinal cord and the splanchnic nerves.
2. Messages sensed by receptors in the intrinsic plexuses which then travel to the brain by way of the vagal afferent nerves to initiate efferent inhibition through vagal or splanchnic nerves.
3. Messages initiated in the intrinsic plexuses which pass through splanchnic afferents to the spinal cord and return by way of splanchnic efferents. These latter constitute intestino-intestinal reflex inhibition.

Since both stimulatory and inhibitory effects on the gut can be initiated either in the gut itself or in the central nervous system, there is reason to term intestinal sensing of its own motor activity a "visceral brain." This concept has gained support from recent findings that many neurotransmitters are common to both the nerves of the gut and those of the brain, in particular, the polypeptide transmitters. The visceral brain is able to excite or depress motor activity either through the intrinsic plexuses directly, through spinal cord reflexes or through activation of the brain. Furthermore, both excitatory and inhibitory messages are carried by both the parasympathetic and sympatheic portions of the autonomic nervous system. The neurotransmitter agents involved in these processes include acetylcholine, catecholamines, polypeptides, serotonin and possibly purine nucleotides. Inhibition of motor activity can occur because the motor unit is depressed by a particular transmitter, because a transmitter blocks transmission at a ganglion or because of a central effect.

As an example of the coordinated neural regulation of motor activity, let us consider the responses taking place in the stomach following eating of a meal. Complex propulsive and mixing actions occur during 1) relaxation of the upper portion of the stomach to accommodate to the volume of food that has just been swallowed, and 2) with organized contractions and peristaltic activity ongoing in the distal half of the stomach as this organ empties. Simultaneously gastric juice is being secreted from parietal, peptic and mucous cells, and the hormone gastrin is being released into the blood from the G cells. Impulses emanating in the cerebral cortex, hypothalamus, globus and limbic system related to the perceptions of hunger and eating are relayed to the vagal nuclei, from which impulses will be transmitted via the vagi to the stomach. Vagal efferent endings impinge directly upon the effector units, namely visceral smooth muscle cells and secretory cells, causing myoelectric responses leading to contractions and to secretion. The transmitter in this case is acetylcholine. There are also inhibitory efferent fibers which travel in the vagi to the smooth muscle cells of the proximal stomach. The transmitter is probably a polypeptide, such as vasoactive intestinal peptide, or a purine nucleotide, such as ATP. In addition, postganglionic sympathetic nerves release catecholamines which inhibit motor activity in visceral smooth muscle. Vagal efferent terminals synapse with the intrinsic nerve plexuses within the wall of the stomach. The intrinsic nerves are stimulated to release excitatory transmitters, such as acetylcholine, and inhibitory transmitters as well. The intrinsic input reinforces and sustains appropriate contraction and relaxation of the muscle. As the food enters the stomach and stretches the wall, receptors activate the intrinsic plexuses and relay afferent messages through the splanchnic nerves to the cord and through the vagi to the brain stem, further adding to the nerve traffic going to and from the stomach. The response of particular areas of the wall of the stomach to nervous input depends not only on the particular transmitter which is released, but also upon the kinds of receptors on the muscle cell and the interaction between different kinds of nerve endings on the motor unit.

Effects of Vagotomy

The most common operation performed for the treatment of peptic ulcers in the United States involves sectioning of the vagus nerves below the heart. The procedure can be accomplished by a transthoracic or abdominal approach and, until recently, consisted of total section of both vagus nerves. Because of the undesired effects of total section, recent developments have included more selective sections in which only those branches of the vagus nerves which pass to the secretory portions of the stomach are cut, thereby avoiding section of the nerves which pass to other gastrointestinal organs. Total bilateral vagotomy not only causes reduction in the rate of gastric secretion of acid (which is the intended effect) but also causes reduction in the secretion of gastrin, pancreatic enzymes and hepatic bile. In addition, there are inhibitory effects of vagotomy on gastrointestinal motility. These include delayed emptying of the stomach, slowed peristalsis of the esophagus and reduced contractile activity of the small and large intestine. Since intestinal contractions not only propel the contents but also act as a brake against excessive rates of movement of fluid through the gut, it is not surprising that one of the major side effects of vagotomy in patients is chronic diarrhea. Since the digestive and absorptive processes of the gastrointestinal tract depend on coordinated motility and the ready availability of various digestive juices, section of the vagi just below the diaphragm is not an altogether beneficial process. Hopefully the selective vagal section will remedy these adverse effects of the more extensive operation.

Summary

There are three major types of nerves in the gastrointestinal tract, namely extrinsic autonomic nerves which are visceral efferent fibers from the sympathetic and parasympathetic systems, intrinsic neurons organized in plexuses within the walls of gastrointestinal organs and afferent nerves from the viscera which travel to other nerves located either in the plexuses, ganglia, spinal cord or the brain stem. Activity of these nerves can initiate, modify or inhibit gastrointestinal exocrine secretion, endocrine secretion of hormones and gastrointestinal motility. Extrinsic and intrinsic parasympathetic efferent nervous input usually causes an increase in salivary secretion, gastric secretion of both acid and pepsin, pancreatic secretion of enzymes, antral release of gastrin into the blood and enhanced motility in several gastrointestinal organs. The section of the major nerve trunks carrying these fibers, as occurs in bilateral vagotomy, causes disruption of secretory and motor events in much of the gastrointestinal tract.

chapter 7

Motility

how mechanical movements of gastrointestinal organs occur, are regulated and contribute to gastrointestinal functions

Under normal conditions food and drink are moved from the mouth to the intestine, digested there after mixture with gastrointestinal juices, and the unabsorbed materials are propelled along the gut to the anus for excretion. This process of moving materials some 450 cm takes place in a regulated fashion and depends upon the intrinsic ability of the gastrointestinal tract to display mechanical motions. These movements take many forms and are termed "motility."

A factor which influences both the onset and the nature of gastrointestinal motility is the spontaneous electrophysiological activity of visceral myogenic cells. Gastrointestinal smooth muscle is an excitable tissue whose cell membranes actively generate electrochemical gradients that are responsible for storage and discharge of electrical currents, for separation and fluxes of ions and for the existence of the muscle in either a relaxed or a contracted state. Although visceral smooth muscle has not been studied as intensively as striated muscle or cardiac muscle, a reasonable assumption is that most of the principles of muscle physiology apply to the myogenic elements of the digestive organs as well. Mechanical activity of gastrointestinal organs depends upon electrochemical events occurring in the membranes of individual muscle cells. Motility is a coordinated function which depends additionally on the influence imposed by one cell upon another. This coordination is effected through electrical messages traveling from one portion of the muscular wall to the next, thereby providing a basic electrical rhythmicity to the mechanical events, much in the manner of a series of coupled oscillators.

The motions of our hollow gastrointestinal organs occur, for the most part, without our awareness. These involuntary actions are influenced greatly by autonomic nerves, especially parasympathetic motor fibers distributed by the vagi. Parasympathetic stimulation releases the cholinergic transmitter or neurocrine agent, acetylcholine, which initiates or augments propulsive movement down the esophagus, stomach, small intestine and colon. At the junctions between these organs the circular smooth muscle ring behaves like a valve or sphincter impeding propulsion of material from one organ to another. Cholinergic stimulation also relaxes these sphincters allowing contents to be propelled from organ to organ. There are, in addition, extensive networks of postganglionic intrinsic nerves located throughout the walls of these hollow organs. Extrinsic parasympathetic preganglionic nerve fibers synapse with the intrinsic nerves, most of which release acetylcholine. Intrinsic nerves may also be stimulated mechanically, as during stretch of the wall of a hollow organ by swallowed food. There are, however, some intrinsic nerves which inhibit contraction, presumably through release of a neurocrine agent other than acetylcholine. Such relaxation permits receptive accommodation of food in a distal segment of the hollow organ as the proximal segment is contracting behind the bolus to propel the food.

Sympathetic postganglionic fibers emerge from the celiac, mesenteric and hypogastric ganglia to accompany splanchnic arteries coursing to the viscera. Sympathetic stimulation releases adrenergic neurocrine agents, the catecholamines, norepinephrine (mainly) and epinephrine, which inhibit propulsive movements and do not relax the sphincters.

Recent expansion of our knowledge about gastrointestinal hormones indicates that these endocrine agents—at least at high concentrations—alter the force and frequency of propulsive movement as well as the state of relaxation of the sphincters. Four of these hormones are established, physiologically important endocrine regulators of gastric or pancreatic secretion. Gastrin and cholecystokinin (CCK) generally stimulate propulsive motion, while secretin and gastric inhibitory peptide (GIP) inhibit such activity. There are other recently identified endocrine agents which affect motility. In addition, tissue substances or paracrine agents, such as prostaglandins, 5-hydroxytryptamine and bradykinin, elicit motor responses from gastrointestinal smooth muscle.

Finally, it seems likely that the aforementioned neurocrine, endocrine and paracrine agents stimulate muscle cell membrane receptors which act, in turn, through common metabolic intermediaries in the smooth muscle cell. These intermediaries may include membrane enzymes ($Na^{+}+K^{+}$–ATPase, its phosphorylating enzyme, adenylate cyclase, guanylate cyclase, etc.), their substrates and products (ATP, cyclic AMP, cyclic GMP, etc.), the ions related to their function (Na^{+}, K^{+}, Ca^{++}, Mg^{++}, etc.), and the ultimate macromolecular targets of membrane metabolism (troponin, actin, myosin, etc.). It is beyond the scope of this chapter to review recent advances in general muscle metabolism, and we are missing many elements in the current picture of this subject as it applies to gastrointestinal motility.

A summary of excitatory and inhibitory factors involved in gastrointestinal propulsion is depicted in Figure 7.1.

The remainder of this chapter is arranged according to the gastrointestinal organs which exhibit motor function. As such the text becomes a guided tour of motility in the hollow organs beginning with the mouth and proceeding aborally to the anus.

Mouth and Pharynx

Chewing of food is considered to be a voluntary act, although chewing responses to food may occur reflexly in decerebrate patients. The mandible is moved up and down, forward and sideways during chewing to permit the teeth to tear and grind the food. Chewing elicits afferent messages to the central nervous system which, in turn, initiates stimulatory output to the salivary glands, stomach, pancreas and liver via neural or neuroendocrine pathways. Chewing also enhances the opportunity for food to tantalize the taste buds and probably permits a person to reach satiety earlier than would occur if the food were bolted without chewing. Nevertheless, swallowing unchewed or slightly chewed food is common in our society with its hurried meals, and lack of chewing does not seem to be related to maldigestion or malabsorptions syndromes.

Swallowing of food is initiated at will but once started proceeds reflexly. The act involves both oral and pharyngeal structures. Food—masticated or otherwise—is lifted by the tongue against the palate. The bolus of food forces the soft palate upward sealing off the nasopharynx as a possible route for the food. The posterior tongue pushes the bolus backward using the hyoid bone as a fulcrum and the hard palate as a trough along which the tongue is able to guide the food. As the bolus touches the posterior

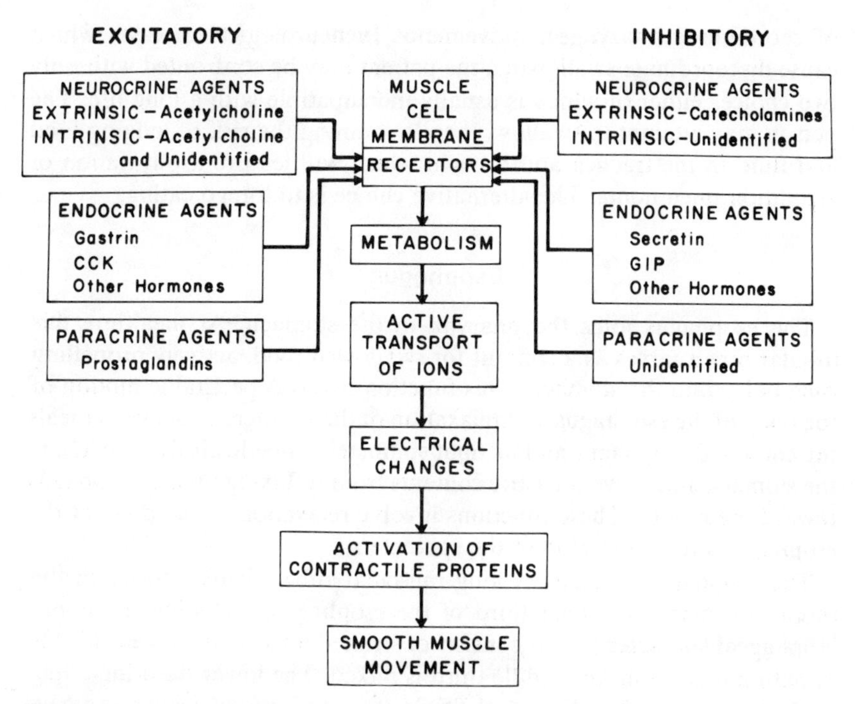

Figure 7.1. Excitatory and inhibitory factors regulating gastrointestinal smooth muscle movement.

wall of the pharynx, the pharyngeal musculature contracts in a coordinated and rapid manner to enact swallowing. First, the upper pharyngeal muscles contract above the bolus forcing it down the pharyngeal channel away from the nasopharynx. Below the bolus the pharyngeal muscles relax to make way for the food. Relaxation of the cricopharyngeus muscle and upward movement of the cricoid cartilage opens the upper esophageal sphincter giving access to the esophagus. Simultaneously the epiglottis moves up and the glottis closes off the trachea to the food. As the bolus passes downward, each portion of pharyngeal muscle contracts behind to apply continuous thrust to the food and forces it into the esophagus. This coordinated wave-like motion of hollow organs is termed "peristalsis." Once the bolus passes the sphincter, each pharyngeal muscular structure returns to its resting position and condition to await another swallow. The complex pharngeal sequence of swallowing is accomplished in 1 second.

Coordination of swallowing occurs centrally in an area of the reticular formation. Destruction of this swallowing center deprives the individual of coordinated pharyngeal movements. In neurological diseases which cause discoordinate swallowing, the patient may be confronted with only two choices either of which is usually incompatible with a long life. The person may attempt to swallow, thereby running the risk of lodging food and fluid in the trachea and bronchi which will lead to asphyxiation or aspiration pneumonia. The alternative choice is to forego eating.

Esophagus

The esophagus joins the pharynx to the stomach. At mealtime this tubular organ serves as a conduit for swallowed food, actively propelling each bolus into the stomach. This function involves peristaltic motion of the body of the esophagus and relaxation of the sphincters. Between meals the collapsed esophagus and its tight sphincters impede air from reaching the stomach and prevent gastric contents from refluxing up the esophagus toward the mouth. These functions involve relaxation of the body of the esophagus and contraction of the sphincters.

The esophagus is a 20-cm long muscular tube located mostly in the thorax. In man, the upper third of the esophagus, including the upper esophageal sphincter (UES) consists of striated muscle; the lower third is smooth muscle; and the middle third is mixed. The lower third including the lower esophageal sphincter (LES) is organized into two layers, namely an outer layer in which the muscle is arranged along a longitudinal axis and an inner layer in which the muscle is ordered along a circular axis. Except at mealtime, pressures within the lumen of the esophagus correspond to intrathoracic pressures, *i.e.*, a few millimeters of mercury below atmospheric. When esophageal muscle contracts during propulsive motions, pressures can increase several-fold. Between meals, pressures generated by the two sphincters are at least 10 mm Hg, thereby sealing the cavity of the esophagus from the stomach and pharynx.

There is also neuroendocrine coordination of esophageal motility with activity in nearby organs that permits effective esophageal functioning. Thus, for example, during inspiration reflex contraction of the cricopharyngeus muscle occurs, thereby closing the UES to the inflow of air. Similarly, closure of the LES is stimulated by release of hormones from the stomach and upper gut after food has passed through the esophagus to reach those organs.

Esophageal propulsion of a bolus of swallowed food takes the form of peristaltic movement. Peristalsis consists of a coordinated series of muscular contractions and relaxations. As the food enters the upper esophagus after swallowing, the bolus stretches the walls triggering both local and central reflexes which initiate peristalsis. Initially, a segment of upper esophageal circular muscle contracts to form an indentation in the wall of the esophagus. Then the longitunal muscle of this segment contracts, and the indentation seems to move a short distance along the esophagus. Next there is a contraction of the circular muscle of the adjacent segment of esophagus as the original segment relaxes, followed by contraction of the longitudinal muscle of the second segment. The propagation of the indentation in the wall is smoothly engineered and gives every appearance of a wave of contraction flowing along the length of the organ. A peristaltic wave traverses the esophagus in 10 seconds. Since muscle contraction increases intraluminal pressure behind the bolus, food is pushed aborally toward the LES. Liquids can move along the esophagus more rapidly than peristaltic progression. Secondary, peristaltic waves may also be initiated in the absence of swallowing and serve to move retained food or refluxed gastric contents out of the esophagus.

Esophageal peristalsis can be initiated by stretching the walls of the organ in a vagally denervated esophagus, indicating the important role of intrinsic cholinergic nerves. These local nerves which release acetylcholine are distributed throughout the muscular coat of the organ and can also respond either to impulses from adjacent neurons in the intrinsic plexus or to vagal stimulation. Furthermore, muscle cells can be stimulated directly by vagal nerve fibers. Therefore, there is dual neural regulation of muscle activity, as well as two pathways for activating muscle distal to a peristaltic wave: intrinsic nerve plexuses and extrinsic vagal innervation (Figure 7.2). Swallowed food stretches the upper esophagus sending an afferent message to central vagal centers which relay impulses along vagal motor nerves to the upper esophageal muscle and to lower portions as well. Stretch of the wall also stimulates intrinsic neuronal activity. With stimulation of the upper esophagus preparatory messages are sent distally via the network of interconnecting intrinsic nerves. This feedback neural regulation is repeated as the bolus of food progresses, stretching each successive segment of esophagus in its path. The result is a finely regulated myoneural effect—peristalsis.

Myogenic electrical activity of the esophagus can be detected in the form of depolarizations or spike potentials emanating from muscle cells in response to acetylcholine. Following stretch of the upper esophagus by a bolus, spike potentials are recorded moving down the length of the

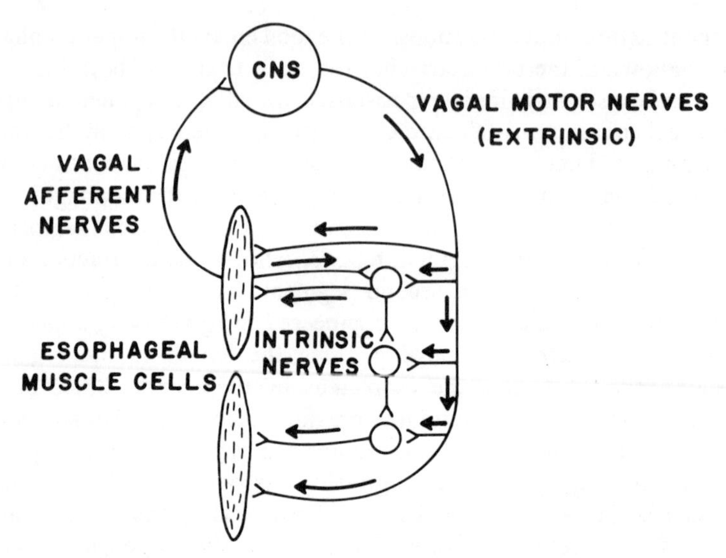

Figure 7.2. Regulation of esophageal muscle activity by extrinsic and intrinsic nerves.

organ at a rate several times faster than the peristaltic wave. This suggests that the lower esophagus requires some form of advance notice to respond smoothly to the coming bolus of food pushed by its peristaltic wave.

At the junction of esophagus and stomach the lining muscle of the distal 2 cm of esophagus is indistinguishable from the muscle above it. Nevertheless, this region, the LES, is a clinically distinguishable structure, since dysfunction of the sphincter causes one of the most common symptoms of which patients complain—heartburn. At least two serious diseases are attributable to chronic dysfunction of the LES, namely, reflux esophagitis and achalasia.

At rest the LES is closed, and its constricted state appears to be a characteristic of this unique muscle. As the bolus of food approaches, cholinergic extrinsic and intrinsic nerves release acetylcholine at the sphincteric muscle whose response is to relax. This response permits passage of the food into the stomach.

Clinically, one can assess LES tone by measuring intraluminal pressure at the LES in patients (normally around 10 mm Hg). When LES pressure is reduced, indicative of a relaxed sphincter, food can travel from the esophagus into the stomach, which is a normal event. Gastric contents, unfortunately, can also reflux from the stomach into the lower esophagus and gastric juice is corrosive to the esophageal mucosa. If this reflux occurs excessively the mucosa becomes inflamed; the disease is termed "reflux esophagitis." Its striking symptom is heartburn, which consists of a burning or squeezing retrosternal pain radiating sometimes to the neck. A clinical test to determine whether esophagitis is present is to infuse an acidic solution into the lower esophagus to elicit heartburn in the susceptible patient. As the name of the symptom suggests, the physician may have need to distinguish between heartburn referable to esophagitis and pain caused by ischemic heart disease. LES pressure is decreased by several foods which are associated with heartburn in susceptible subjects: coffee, fats, chocolate, peppermint, citrus juices and ethanol.

Among the endogenous factors in the body which lower LES pressure are endocrine agents (secretin, GIP, vasoactive intestinal peptide, glucagon, CCK), neurocrine agents (acetylcholine, β-adrenergic stimulation, dopamine) and paracrine agents (E and A type prostaglandins). Accumulation of cyclic AMP in the LES also leads to its relaxation.

Conversely, the gastrointestinal hormones gastrin and motilin evoke increased LES pressure and prompt closure of the valve. It is tempting to speculate that the release of gastrin that occurs at the time of feeding maintains constriction of the LES which is relieved only at those times when a bolus of food approaches. Enhanced LES constriction would impede reflux of gastric juice whose secretion is increased at the time of feeding.

Achalasia is a disease in which LES tone is excessively great due to lack of inhibitory fibers in the intramural plexus and the sphincter fails to open easily at the time of swallowing food. If the food passes, the patient with achalasia may experience discomfort or pain and will regurgitate. Pain of achalasia can also be confused with the pain of ischemic heart disease, and achalasia leads to weight loss. Treatment of achalasia may require surgery.

Stomach

The bag-shaped stomach stores swallowed food and fluid for minutes to hours. During storage a small amount of digestion occurs, since the stomach secretes a proteolytic enzyme, pepsin. In addition, soluble foods dissolve in the stomach, and gastric juice will bring the ingested material close to the osmolality of plasma. Some bacteriocidal action occurs in the stomach, since actively secreted gastric juice has a pH of about 1.0. The long storage time of food in the stomach caused by the slow emptying of gastric contents also guarantees presentation of limited volumes of material to the upper intestine. The storage capacity of the duodenum is low and its response to sudden stretch provokes unpleasant symptoms such as nausea, abdominal discomfort, vomiting and faintness. Loss of the storage function of the stomach occurs in patients who have undergone surgical removal of the stomach and occurrence of the aforementioned symptoms after eating is termed "dumping syndrome."

The muscular wall of the stomach is innervated by the vagi. Acetylcholine stimulates contraction of smooth muscle elements within the wall, causes relaxation of the pyloric sphincter and activates intrinsic nerves of the myenteric plexus. Postganglionic sympathetic fibers from the celiac ganglion release catecholamines which inhibit gastric smooth muscle. The vagi also carry inhibitory fibers which release an unidentified neurotransmitter. Stretch of the walls of the stomach initiates afferent impulses through both the parasympathetic and sympathetic extrinsic nerves. Such long reflexes modulate gastric motility through the aforementioned efferent pathways either initiating, enhancing or inhibiting movements of the stomach. Stretching of the gastric walls also activates the intrinsic cholinergic nerves which constitute a short reflex.

In considering this multicomponent nervous regulation which can either stimulate or inhibit gastric smooth muscle, it must be kept in mind that the stomach is required to accommodate to an ingested meal in two ways: the organ must relax to receive a volume sometimes exceeding a liter in a matter of minutes, and the organ must initiate and maintain coordinated peristaltic activity to mix the contents and slowly empty itself. Both relaxation of the stomach and increased motor activity of its wall occur simultaneously after a meal is eaten. In order to better carry out these dual actions (storage and mixing of the meal), there is anatomic localization of gastric motor activity. Thus, the upper half or proximal stomach is the part that relaxes to receive the meal and the lower half or distal stomach is the part that mixes and empties the meal.

Within the proximal stomach pressure is about the same as intra-abdominal pressure. If a thin, fluid-filled nasogastric tube is swallowed by a patient, the physician can determine from changes in recorded pressures where the tip of the tube is located. If the tip is in the esophagus and the patient inhales, pressure will decrease; if the tip is in the proximal stomach, inhalation will be accompanied by a pressure rise.

Pressure in the lumen of the proximal stomach rises only slightly when a meal enters because of the relaxation of the walls of the organ. Some relaxation of the proximal stomach occurs with each swallow, permitting the organ to anticipate arrival of food which may not reach the stomach for several seconds. Reflex relaxation of the proximal stomach associated with the act of swallowing is mediated by vagal inhibitory fibers. Following vagotomy distensibility of the proximal stomach is decreased. Distensibility of the proximal stomach is increased by large doses of gastrointestinal hormones, gastrin and CCK. At some time after entry of a meal, the stomach will begin emptying its contents slowly. The proximal stomach contributes to this process by generating a little motor activity and by becoming less distensible, thereby providing a little pressure behind the meal that must be emptied.

The lower half or distal stomach is considerably more motile than the proximal part. Smooth muscle cells of the outer longitudinal layer of the wall of the distal stomach possess a membrane which depolarizes and repolarizes rhythmically. This electrical activity is termed the "basic electrical rhythm" (BER) or slow wave. The BER is present whether or not there are any mechanical contractions of the wall (Figure 7.3). In patients the BER has a frequency of 3 to 5 cycles/min anywhere in the distal stomach. Despite the uniform frequencies of the BER, there is a phase lag from the midportion of the greater curvature of the stomach to the pylorus. Consequently, not all parts of the distal stomach depolarize simultaneously. The relationship of the slow waves to mechanical con-

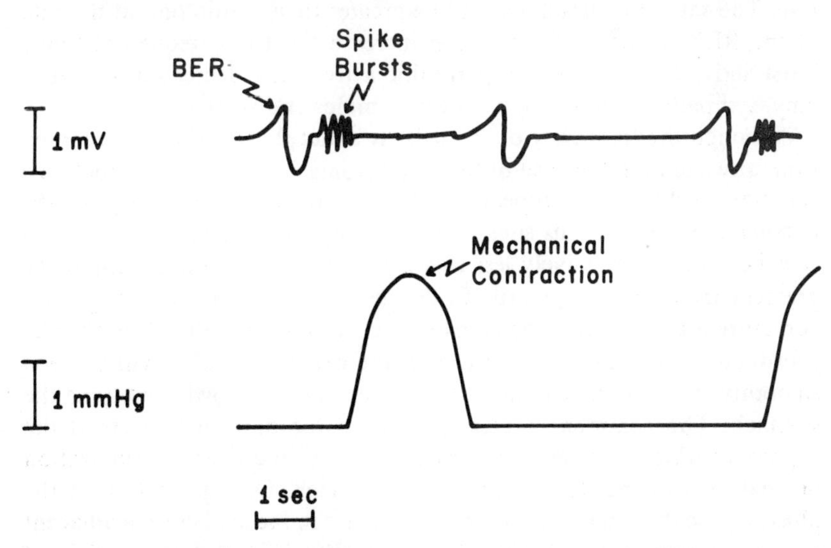

Figure 7.3. Relationship between electrical activity (BER and spike bursts) and mechanical contraction in the distal stomach.

tractions is complex and will be discussed shortly; however, it should be noted at this point that peristaltic waves also originate at the midpoint of the greater curvature and travel along the path set by the BER to the pylorus.

Another event observed from recordings of electrical activity in the wall of the distal stomach is a rapid depolarization of muscle membrane. This occurs as a series of several bursts appearing during the repolarization phase of a slow wave (Figure 7.3). Not every slow wave, however, has the spike bursts.

Mechanical contractions of gastric smooth muscle bear a definite relation to both the BER and to the spike bursts (Figure 7.3). A contraction occurs only when preceded by spike bursts. Spike bursts appear during the late phase of some but not all slow waves of the BER. Consequently, those slow waves which are not accompanied by spike bursts are also not followed by a muscular contraction, although each contraction is preceded by a slow wave upon which is superimposed a set of spike bursts. Thus, the BER establishes the maximal rate at which peristaltic contractions can occur—3 to 5/min in man.

At rest the rate of spike burst activity and of contractions is about 3/min. With vagal stimulation or gastrin stimulation the rate of spike burst firing and of peristaltic contractions increases but cannot exceed the BER rate. The rate of contractions will be greater than 5/min only if the rate of the BER increases. Gastrin can increase the BER frequency. Spike burst activity and, therefore, peristaltic activity are decreased by vagotomy, sympathetic stimulation, catecholamines and secretin.

Although the frequency of the BER is about the same when recorded from any site in the muscle of the distal stomach, not all sites depolarize simultaneously. There appears to be an organized phase lag in the depolarization of various sites such that the phase lag between adjacent sites becomes progressively less as we proceed from the midpoint of the greater curvature to the pylorus. Consequently, the midpoint of the greater curvature is considered to behave like the pacemaker of the stomach. The coordination of phase lag assures that the peristaltic waves will proceed smoothly from greater curvature to pylorus, thereby slowly emptying the stomach. The common surgical procedure for duodenal ulcers is the vagotomy which interferes, unfortunately, with the overall organization of phase lag in the BER of the distal stomach. Disorganization of the phase lag, so that there is no longer a decrease in the lag between adjacent muscles as we approach the pylorus, results in a discoordination of peristaltic activity and impaired emptying of the stomach. Additional surgery (pyloroplasty) is performed with the vagotomy to compensate for the damage done to the mechanism of gastric emptying.

Emptying of the Stomach

Gastric emptying is a complex function because two requirements must be satisfied; namely, the stomach must empty ingested food and fluid as well as its own secretions over a reasonable period of time, but the stomach must not empty too rapidly. Complete emptying of a meal with a volume, let us say, of ½ liter of solids and fluid may consume 2 hours during which time several hundred peristaltic waves may have developed and moved toward the pylorus. Each of these waves caused only a few milliliters of gastric contents to leave the stomach and enter the duodenum.

The factors which influence emptying of food from the stomach include the presence of an ingested load in the organ, the composition of the gastric contents, the level of motor activity in the proximal and distal portions of the stomach, tension at the pyloric sphincter, duodenal motor activity, the extrinsic and intrinsic parasympathetic nerves and the gastrointestinal hormones released during and after feeding—gastrin, secretin, CCK and GIP. The interrelationship between these many factors appears in Figure 7.4.

During resting periods of the stomach (some time after feeding when the individual is not experiencing hunger, for example, at 3 p.m. or 3 a.m. for the average person), motor activity is minimal, secretions are negligible and little emptying is necessary. Entry of a meal into the stomach initiates motor activity which predisposes to emptying. Food stretches the walls of the stomach, thereby triggering short reflexes through the intrinsic cholinergic neurons and long reflexes through the vagal afferent neurons. The end result is release of acetylcholine by efferent pre- and postganglionic fibers in the vicinity of smooth muscle cells of the gastric wall which increases spike burst activity and increases the rate of peristaltic waves. Peptides and amino acids in the food increase G cell elaboration of gastrin into the blood draining the antral mucosa. Food also buffers gastric acid to pH levels above 2.0 which prevents feedback inhibition of gastrin release.

The composition of the gastric contents influences gastric emptying. An important factor here is the osmolality of the fluid in the stomach. Hypotonic and isotonic solutions empty faster than hypertonic. Thus, most of a swallowed glass of tap water will be emptied in less than ½ hour, whereas a milkshake will sit for much longer. Hypertonic solutions attract some fluid from the plasma, lowering the osmolality in the lumen. Beside its osmotic effect, the ingested material will contain specific food products which can diminish the rate of gastric emptying. Specifically, gastric emptying of fat and protein encourage the duodenal mucosa to release CCK and GIP, two hormones which inhibit both gastric secretion and motility. This constitutes negative feedback regulation of emptying.

There is a relationship between the comparative levels of motor activity in the distal stomach versus that of the duodenum. Emptying is accelerated by agents which stimulate gastric peristalsis in excess of duodenal, such as acetylcholine and gastrin, and is decreased by agents which augment duodenal peristalsis more than gastric, such as secretin and CCK. The proximal stomach behaves passively by relaxing to accommodate ingested loads of food and then contracting mildly to apply some constant pressure to propel the contents toward the pylorus. Increased pyloric sphincter tone will impede emptying. Such heightened tension occurs in response

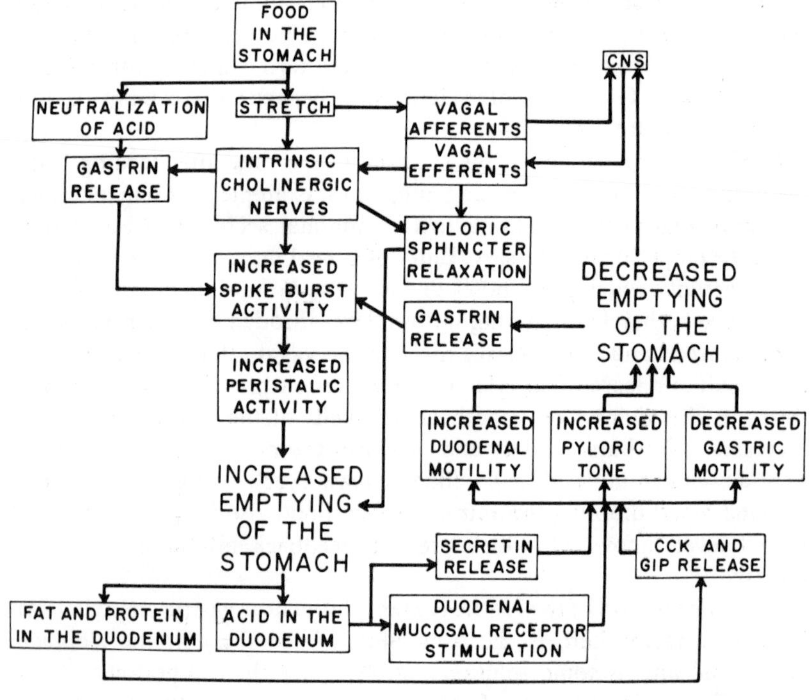

Figure 7.4. The complex control of gastric emptying.

to gastric emptying of its acidic juice which causes duodenal mucosal release of the hormone secretin. Acid also triggers duodenal mucosal nerve receptors which connect reflexly to the sphincter. Both an enterogastric reflex and secretin prompt increased sphincteric tone and slow emptying.

The cholinergic neurotransmitter increases distal stomach peristalsis and relaxes the sphincter, thereby enhancing gastric emptying. Acetylcholine also releases gastrin, which stimulates distal stomach motility and aids emptying. The duodenal hormones secretin (released by acid), CCK and GIP (both released by fat and protein) inhibit gastric emptying by slowing gastric peristalsis, by increasing pyloric tone and by stimulating duodenal motor action.

Given so complex a mechanism for the emptying of the stomach, it is not surprising that disorders of gastric emptying occur and are responsible for hospitalization. Aside from vagotomy, there are two other major causes of interference with emptying of the stomach; namely, stenosis and ulceration of the pyloric channel. Pyloric stenosis usually occurs in infants and is due to congenital muscular overgrowth of the sphincteric muscle. Symptomatically the baby is continually hungry, vomits excessively (because the stomach cannot empty properly) and loses weight. Surgical incision of the sphincter usually relieves the pylorospasm and permits the stomach to empty. Chronic peptic ulcer of the pyloric channel leads to scarring and contraction of the pyloric outlet. With an acute flare-up of the ulcer there is inflammation and edema, which closes the pyloric orifice altogether. If this state persists the stomach is said to be obstructed and the patient is in a life-threatening condition, since failure of the stomach to empty food into the gut imposes starvation on the individual. Treatment of pyloric obstruction requires hospitalization, parenteral alimentation, correction of fluid/electrolyte abnormalities and continuous aspiration of gastric contents coupled with antisecretory drugs to put the stomach to rest until the ulcer becomes quiescent. If these conservative measures fail to relieve the obstruction, surgical intervention becomes necessary.

The stomach regularly empties its contents into the duodenum. There is also an infrequent method of gastric emptying in the wrong direction which we know as vomiting. Nearly all humans have experienced vomiting at least once and the symptom is hardly a cause for alarm, although chronic and frequent vomiting can signal serious disease, *e.g.*, pyloric obstruction or achalasia. Usually a person becomes aware of an incipient need to vomit, because of a subjective sensation called nausea. Just before the act of vomiting starts there is a diffuse autonomic discharge in the body. Sympathetic nervous activation causes tachycardia, tachypnea and sweating. Parasympathetic activation evokes relaxation of LES and UES, increased gastric motility and brisk salivation. At this stage the person usually inhales deeply, thereby moving the diaphragm and stomach down into the abdomen, closes his glottis, and spasmodically contracts the anterior abdominal muscles (retching) which squashes the stomach up against the diaphragm. Gastric contents are propelled rapidly up the esophagus and pharynx and are ejected from the mouth.

Small Intestine

Several forms of motility are evident in the small intestine. This organ is the major site of digestion and absorption of nutrients. It requires both mixing of its liquified contents (chyme) with digestive juices and forward propulsion of the chyme to expose digested substances to the absorptive surface. Regular peristaltic activity starts near the entry sites of the common bile duct and main pancreatic duct in the proximal duodenum. Not surprisingly, this area initiates the BER and has been viewed as the pacemaker of the small intestine. Both the slow waves and the peristaltic contractions progress from the duodenum down the length of the gut to the ileum at a velocity of 1 cm/min. There are about 10 slow waves per minute on the average, more in the duodenum and less in the ileum.

In addition to propulsive peristaltic waves there are simple contractions of circular muscle which do not progress, although these motions may occur rhythmically and at recurrent sites on the gut. These are called "rhythmic segmentations" and are also noted in the colon. When the intestine is opacified by swallowed barium during x-ray filming, rhythmic segmentation causes the gut to appear like a string of sausage links. Rhythmic segmentation serves mainly to mix the chyme with digestive enzymes from the pancreas and to assure exposure of all parts of the chyme to the mucosal surface. Differing rates of segmentation along the intestine also facilitate propulsion of the chyme.

The mucosal lining is not smooth because of the ridges formed by the valves of Kerckring. These humps in the wall contain motile smooth muscle which causes the surface to undulate, thereby further stirring the chyme and exposing digested food to the absorptive surface. The surface projections studding the mucosa are finger-like villi which contain smooth muscle and move in the stream of chyme like the arms of a sea anemone. These motions also mix the chyme with digestive fluids and present added surfaces for absorption. As a result of this combination of slow propulsion of small amounts of nutrient with continuous mixing and exposure of the chyme to the vast surface of the gut, digestion and absorption are remarkably successful processes. In the end 100% of digestible dietary carbohydrate, protein and fat and over 90% of the water and salt are absorbed from the lumen of the small intestine into the blood daily. In the absence of normal motility, brought about by peritonitis, a condition termed "adynamic ileus" occurs. Much fluid accumulates in the lumen of the gut and normal alimentation is impaired.

The BER sets the maximal rate at which peristaltic contractions will occur in the small intestine. In the duodenum the rate of the BER is about 12 cycles/min with a decrease to 8 cycles/min in the distal ileum. There is a phase lag in the appearance of slow waves indicating dominance of proximal muscle over adjacent distal muscle in the manner of coupled oscillators. Slow waves appear in the human upper small intestine every 5 seconds on the average at a given site and muscular contractions appear, consequently, at a maximal frequency of once every 5 seconds at the same site. Peristaltic contractions occur only if spike bursts have taken place during the repolarization phase of the preceding slow wave.

Vagal efferent fibers innervate the small intestine, stimulate spike burst activity and convert the quiescent gut into a more motile organ. This nervous influence occurs normally at meal times and also in response to emotional stress. The intrinsic cholinergic nerves of the myenteric plexus are stimulated by vagal efferents and by stretch of the wall of the hollow organ. Such stimulation, in turn, releases acetylcholine in the vicinity of smooth muscle cells and evokes spike bursts and contractions. Congenital absence of intrinsic inhibitory neurons in a short segment of bowel occurs in Hirschsprung's disease. The involved segment is constricted, and there is failure of intestinal propulsion at the diseased site with great accumulation of material proximally. This disorder is easily corrected by surgical excision of the deficient segment with anastomosis of the two cut ends of normal bowel.

Vagal afferent fibers convey responses to stretch of a segment of gut and cause inhibition of other areas of the intestine. Sympathetic nerves also inhibit small intestinal motility. Small intestinal motor performance is stimulated by endogenous paracrine or endocrine substances released during disease states, namely certain tumors which release serotonin (carcinoid), prostaglandins (thyroid medullary tumor) and gastrin (Zollinger-Ellison syndrome). Activity is also enhanced by CCK and insulin and is inhibited by secretin and glucagon.

Once material is propelled to the ileocecal junction by peristaltic waves, stretch of the distal ileum initiates a local reflex which relaxes the sphincter and allows fluid to squirt into the cecum. Stretch of the cecum initiates a second reflex which closes the ileocecal junction. The time required for a nonabsorbable marker to travel from the duodenum to the cecum is about 5 hours.

Biliary Ducts and Gallbladder

The biliary tree functions in two ways: 1) to convey bile secreted within the liver to the duodenum for participation in the digestion of fat; and 2) to store the watery bile long enough to convert it into a viscid fluid. The structures necessary for providing both the conduit and the reservoir include the hepatic, common bile and cystic ducts, the gallbladder and the sphincter of Oddi. Secretory pressure generated in hepatocytes propels newly formed bile into the ducts. Usually the sphincter is closed and the gallbladder relaxes to accept the bile. Transport of sodium and chloride by the mucosa of the gallbladder results in absorption of much of the water from the bile leaving behind a smaller volume of dark colored, syrupy fluid.

At mealtime the gallbladder contracts and the sphincter of Oddi relaxes to allow extrusion of bile into the duodenum. One principal organic component of bile, the bile salts, behave as detergents emulsifying water-insoluble dietary lipids. Bile salts are essential to the formation of micelles of digested fats which are physical conglomerates of molecules necessary for absorption. The major stimulus responsible for contraction of the gallbladder is the hormone CCK. CCK is released into the blood by cells within the mucosa of the upper small intestine in response to fat and essential amino acids in the lumen of the gut. CCK also relaxes the sphincter of Oddi. Vagal stimulation has effects similar to those of CCK. The rate of entry of bile into the gut also depends upon the contractile state of the duodenum, that is, the bile is squirted into the lumen of the gut between contractions of the duodenum.

Colon

The colon lacks a continuous layer of longitudinal muscle. Its longitudinal muscle is organized into three flat bands called the "teniae coli" which contribute to the division of the colon into pocket-like haustrations. Perhaps as a result of this unusual structure one does not observe much peristaltic activity in the colon from the cecum through the descending colon. The sigmoid colon and rectum do, however, exhibit active peristalsis. Rhythmic segmentation predominates in the colon. As a consequence of this lack of regular propulsive movement the colon is able to store its liquid contents for up to several days while the material is desiccated and converted to semisolid feces. Most of the water and ions delivered from the small intestine are absorbed in the colon.

Eventually the most distal portion of the fecal material in the cecum is squeezed off the main mass and moved fairly rapidly up the ascending colon, along the transverse colon and down the descending colon to the sigmoid. This process is called "mass movement." It appears to be a coordinated colonic motor effort in which proximal wall tension exceeds distal tension sufficiently to push a semisolid mass a distance of 75 cm in a few minutes. Mass movement occurs once or twice daily in the average person, often happening after a meal. This is the "gastrocolic reflex" probably mediated by gastrin release.

Extrinsic cholinergic nerves reach the wall of the colon from two major sources: the vagi innervate the proximal half of the colon and the pelvic nerves distribute parasympathetic fibers to the rest of the organ. These nerves stimulate mechanical activity of the colon. Intrinsic nerves appear to relax colonic muscle since their absence (Hirschsprung's disease) is associated with tonic spasm of the musculature. Stretch of one portion of the colon inhibits other parts of the organ. Emotional stress is reflected in many people by excessive colonic activity with distressing symptoms such as crampy abdominal pain, frequent flatulations, constipation or frequent defecation. This disorder is termed the "irritable bowel syndrome."

The hormone gastrin may be a stimulator of mass movement. Epinephrine and prostaglandins inhibit rhythmic segmentation, but the latter agents also stimulate propulsive activity of the colon. Not surprisingly, the major side effect of prostaglandin-containing drugs is diarrhea.

Defecation is the last major motor function of the enteric canal. Delivery of feces to the sigmoid stretches the wall and initiates peristaltic activity propelling the feces to the rectum whose upper portion relaxes to permit entry of the load. Rectal stretch is signalled to the cerebral cortex to make the individual aware of the forthcoming event. The seated or squatting position is assumed which provides mechanical advantage to the abdomen against which the thighs are abutted. Usually the person inhales, closes the glottis and exhales, thus raising intrathoracic pressure considerably. This "Valsalva maneuver" can stop venous return and lead to cardiac arrest, especially in the elderly. This increase in pressure is transmitted to the peritoneal space and impinges on the serosal surfaces of the sigmoid and rectum. Simultaneously the person contracts the anterior abdominal muscles and further increases pressure within the abdominal cavity squeezing the outside of the viscus holding the feces. The internal anal sphincter relaxes in response to cholinergic nerve stimulation triggered by stretch of the rectum. Cholinergic impulses also stimulate peristalsis and cause extensive longitudinal muscle contraction in the rectum, thereby abbreviating the organ. Owing to the body position the gluteal muscles stretch the external anal orifice. At this time the individual voluntarily relaxes the external anal sphincter removing the last impediment to extrusion of feces from the body.

Given so complex an act as normal defecation it is understandable why constipation is so common a complaint in our population. Some decline in colonic motor function may be involved since the symptom grows in frequency as people age. The common response to this "change in habit" or "loss of regularity" is to resort to laxatives, enemas or other nostrums. A sounder approach involves an increase in fluid intake, addition of foods containing plant fibers (cereals, legumes, etc.) and reassurance from a physician. In a more natural way plant fiber keeps water in the colon, producing stools which are passed more easily.

Diarrhea has been attributed to hyperactivity of intestinal smooth muscle, especially the colon. This view is unproven and is at variance with findings in cholera, an extreme form of diarrhea in which the gut is hypomotile. Similarly, infusion of water into the upper gut in volumes sufficient to produce diarrhea inhibits contractile activity of the intestine. Furthermore, the contents of the normal colon are not watery; hence its overactivity would not alone cause diarrhea. The best understood forms of diarrhea involve excessive secretion of fluid by the small intestine which overwhelms the normal ability of the bowel to regulate a slow propulsion of material along the gut.

Summary

Mechanical motions of hollow structures in the gastrointestinal tract depend upon the activity of smooth muscle cells contained in the wall of these hollow organs. These motions are termed "gastrointestinal motility" and are essential for the digestion and absorption of food and the elimination of waste. The two major forms of motility are those actions which propel food, fluid and waste products along the length of the gastrointestinal tract from one hollow organ to the next and those motions which serve to mix the contents of the lumen.

Between mealtimes there is some level of motility in the gastrointestinal system. This is heightened at mealtime by nervous stimulation coming from the brain and from intrinsic neurons which are stretched by food moving through the tract, and probably by hormones released at this time.

Swallowing is a motor event which is initiated voluntarily but then proceeds reflexly. Chewed food is moved from the mouth into the upper esophagus. Food entering the esophagus and stretching the upper part of that organ initiates a propulsive motor activity called "peristalsis." This form of motility moves food and fluid rapidly along the length of the esophagus. The lower esophageal sphincter is a muscle thickening at the juncture of the esophagus and stomach. It is constricted, except as food approaches the sphincter, when it relaxes.

The upper part of the stomach accommodates to the ingested food and fluid by relaxing. There is only limited propulsive motion in this part of the organ. On the other hand, the lower part of the stomach exhibits peristalsis which slowly propels the contents out of the stomach into the duodenum. Emptying of the stomach is dependent upon numerous factors including the composition of the food and fluid in the stomach; in particular, the osmolality, protein and fat content, the level of motility in the stomach itself, the degree of tone of the pyloric sphincter, the motor activity in the duodenum, the degree of stimulation by both extrinsic and intrinsic parasympathetic nerves and the gastrointestinal hormones which have been released at mealtime (gastrin, secretin, CCK and GIP).

Both propulsive and mixing forms of motility are found in the small intestine. The liquified contents, consisting of both ingested nutrients and of digestive secretions are termed "chyme." Chyme is mixed continuously in order to expose the extensive surface area of the gut to the nutrients and to expose the nutrients to the digestive enzymes. Chyme is also propelled along the length of the small intestine. After absorption of the nutrients from the chyme in the upper small intestine, the remaining substances are mostly waste products which will have to be eliminated from the colon. In the colon there is a limited propulsion and mixing, since the organ will serve as a reservoir for its fecal contents for some days during which valuable electrolytes and water will be absorbed from the feces prior to elimination.

chapter 8

Salivary Secretion

how digestion gets started

Salivary glands are comprised of numerous, small buccal glands distributed throughout the oral mucosa and pairs of larger parotid, submandibular and sublingual glands located in more specific sites of the mouth. Although they differ somewhat in the nature of protein secreted, all have a similar unit structure. This basic functional unit of salivary secretion has been called the "salivon," because it is somewhat analogous to the capsular-tubular structure of the nephron.

Ionic Composition

Salivary fluid is first formed in blind sacs called "acini" both by filtration of plasma and by active secretion of the anions, bicarbonate and chloride, as depicted in Figure 8.1. Active bicarbonate secretion is aided by the enzyme carbonic anhydrase, which catalyzes the conversion of plasma and cellular carbon dioxide to bicarbonate. This transport occurs at the junction of the acini with an extensive, converging tubular system of ducts. Although plasma protein is not filtered by the acini, specific salivary proteins are actively secreted by inverse pinocytosis, including salivary α-amylase, a digestive enzyme, and mucin, a mucoprotein.

Acinar fluid is variably modified in the converging system of ducts by active transport of the cations, sodium and potassium. Sodium is actively

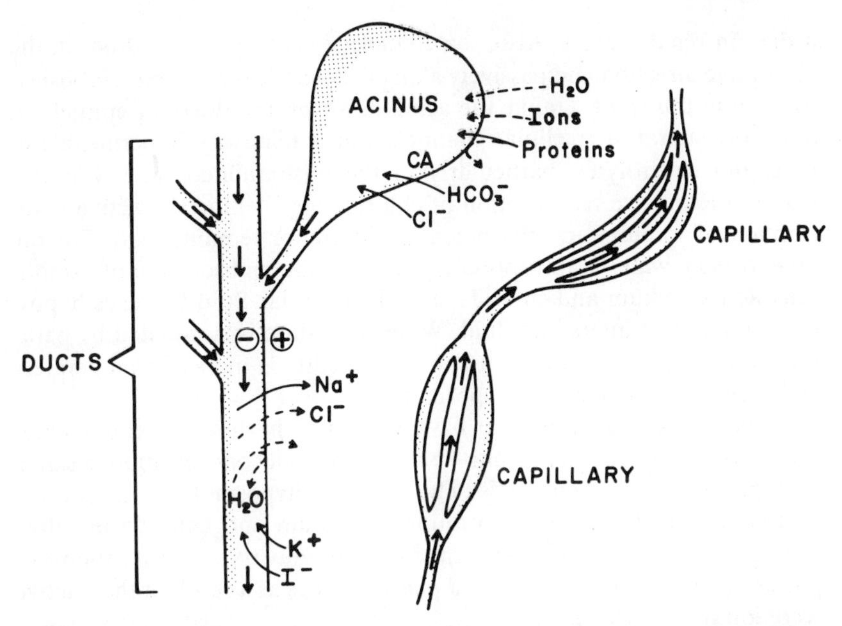

Figure 8.1. Principal movements of fluid and electrolyte in the salivon. Active transport is indicated by *solid arrows*, passive transport by *dashed arrows*, and salivary or blood flow by *heavy arrows*. Ducts form a converging network leading to main ducts. *CA*, carbonic anhydrase.

reabsorbed and potassium actively secreted. At high rates of secretion, ionic modifications in the ducts are less pronounced, because time for cation transport is insufficient. Thus, at high flow rates, the composition of saliva tends to approach that of acinar or interstitial fluid in that ductular transport mechanisms are time-limited.

Ductular ion transport occurs all along the length of the converging ducts and gives rise to an electrical potential difference of up to −70 mV, lumen negative (Figure 8.1). This potential can be abolished by replacing sodium in the acinar fluid with choline or by treating the ducts with ouabain, the typical inhibitor of sodium transport. These observations indicate that the basic ion transport in the ducts is active sodium reabsorption. Active sodium transport out of the duct lumen is largely, but not totally electrogenic; that is, sodium movement is only partly negated electrically by opposite cation or concurrent anion movement. Some active secretion of potassium does occur in partial exchange for sodium, and the hormone aldosterone stimulates this exchange in salivary ducts as it does in renal tubules. Also, some chloride accompanies sodium in the absorptive direction and passively along the electrical gradient established by sodium transport. Unlike the acini, however, the ductular epithelium has short lateral intercellular channels and is relatively impermeable to fluid and electrolytes, particularly in the unstimulated state. For this reason, the passive reabsorption of chloride lags behind the active reabsorption of sodium, and the potential difference remains high. For the same reason, water reabsorption lags behind the reabsorption of osmotically active sodium and chloride, and the ductular fluid becomes hyposmotic relative to interstitial fluid. When salivation is stimulated by parasympathetic impulses, the ductular permeability is increased moderately and the saliva becomes less hyposmotic.

Saliva thus contains the four basic ions of body fluids—sodium, chloride, bicarbonate and potassium—but their overall concentration in saliva is hyposmotic by about 50%. The salivary hyposmolality is entirely attributable to the low concentration of sodium and chloride in saliva resulting from ductular reabsorption. The salivary concentrations of potassium and bicarbonate exceed plasma values as a result of their active secretion in the salivon.

The concentrations of three basic ions increase as salivation is stimulated parasympathetically. Bicarbonate and chloride concentrations increase because the secretory pumps in the acini and proximal ducts are stimulated. Sodium concentration increases because the high flow rates tend to saturate the reabsorptive pump in the ducts. Potassium concentration remains relatively constant, because its ductular secretory pump is not stimulated or inhibited. Such concentration-flow dependence is encountered in virtually all gastrointestinal secretions. Generally, it can be explained in terms of a two-component contribution to the final secretion: an unstimulated component is apparent at low secretory rates and a stimulated component at high rates.

Functionally, the important ionic constituent of saliva is the bicarbonate. At high, stimulated rates of salivation, the bicarbonate concentration rises to twice the plasma value and raises the salivary pH to alkaline values. Salivary bicarbonate and mucin are also effective chemical buffers around neutral pH. These salivary constituents counteract the constant acidification that occurs in the mouth by resident micro-organisms that produce lactic acid. Without neutralization, the bacterial acids would dissolve away tooth enamel, whose mineral composition is an acid-soluble calcium salt.

Finally, iodide is actively secreted in the ducts of the salivon. Salivary iodide concentration can reach levels 10 times that of plasma, being almost completely extracted from arterial blood. Capillary blood flow around the ducts is countercurrent to the salivary flow. This arrangement insures maximal secretion of iodide by minimizing the diffusion gradient against which the pump operates (Figure 8.1). Thus, in the distal duct the arterial iodide is relatively high as is the salivary iodide and in the proximal duct arterial and salivary iodide are both low. Extraction of iodide from salivary arterial blood is complete enough to permit iodide clearance to serve as an estimate of salivary blood flow by the familiar equation:

$$\text{Salivary blood flow} = \frac{\text{(salivary iodide concentration) (salivary flow)}}{\text{(plasma iodide concentration)}}.$$

The function of salivary iodide secretion is presently unknown, but it may relate to the embryological proximity of the salivary glands to the iodide-accumulating thyroid gland.

By replacing hydroxyl in the hydroxyapatite mineral of enamel, fluoride confers resistance to caries on tooth enamel. This action of fluoride is of great importance to general health, because dental caries are the most widespread disease in this country in young people. The incidence of dental caries reaches 90% in many areas, and complete loss of teeth by middle age is common. The edentulous subject can become susceptible to malnutrition. Since the only natural dietary sources of fluoride are tea and sardines, it is essential for most of the population to obtain fluoride through the water supply. Addition of 1 ppm fluoride to water has been demonstrated to reduce greatly the incidence of dental caries in children and loss of teeth in adults. The most effective administration is a fluoride-vitamin preparation for infants before teeth have erupted.

Salivary fluid itself resists formation of caries by cleansing and dissolving food particles from crevices between teeth where bacteria reside. In addition, proteolytic enzymes and thiocyanate present in saliva are bacteriostatic.

There are no commonly encountered human illnesses other than mumps which involve the salivary glands primarily. Nevertheless, there are reports about patients with either a congenital absence of salivary glands or with glands which fail to secrete saliva (aptyalism) because of damage, as after irradiation of the glands during the course of treatment of head and neck cancer, or in Sjögren's syndrome in which neither saliva nor tears are formed. Aptyalism deprives the mouth of the protective features of saliva and buccal structures become vulnerable to pathogenic bacteria. Such patients suffer recurring dental caries and chronic inflammation of the mucosal lining of the mouth (stomatitis).

Protein Composition

Saliva may be classified by its viscosity as thin or serous from the parotid gland and thick or mucous from the sublingual gland. The submandibular gland saliva is intermediate. These categories correlate with the nature of the salivary protein composition. The serous type contains the digestive enzyme α-amylase that splits the internal α-1,4-glucosidic linkages of starch irreversibly to form maltose, maltotriose and dextrins. It requires—and salivary ions provide—a neutral pH and chloride for its optimal activity. Like salivary fluid and electrolyte, α-amylase production is greatly increased by feeding or parasympathetic stimulation. Since food resides in the mouth for only a short interval, most of salivary α-amylase digestion of carbohydrate occurs in the stomach. Salivary digestion ends when gastric acid overcomes swallowed salivary bicarbonate and ingested food buffers to inactivate the enzyme.

The mucous type of saliva contains mucin, a mucoprotein that serves the general function of lubrication. Mucin adheres to food and to the mouth, forming a thin film that reduces friction. It is resistant to digestion by gastrointestinal enzymes and it is a weak buffer. Its synthesis is increased by either sympathetic or parasympathetic stimulation. These stimuli also contract the myoepithelial cells that support and prevent leakage from the acini. Myoepithelial contractions shorten and widen ducts also, leading to the facilitated expulsion of viscous saliva. Additionally, mucous type saliva facilitates passage of the food bolus from mouth to stomach, and saliva generally is conducive to speech.

Stimulation

Unstimulated salivary blood flow and oxygen consumption are 10 to 20 times that of resting muscle. Both salivary blood flow and oxygen consumption can be linearly increased by parasympathetic stimulation or by acetylcholine to 5 times the resting levels. Blood flow to the salivary glands is supportive but not the limiting factor to salivary secretion. In small doses, atropine can inhibit salivary secretion but not the vasodilatation evoked by parasympathetic stimulation.

Prominent salivary vasodilation can be stimulated through release of a salivary enzyme, kallikrein, into the bloodstream. Kallikrein cleaves from a plasma protein an octapeptide called "bradykinin." Bradykinin is an extremely potent vasodilator but does not control salivary secretion. Its exact physiological function is unknown.

The unstimulated rate of salivation is about 0.5 ml/min, principally from the submandibular glands. Salivation can be reduced by dehydration, fear and anxiety. The latter may reflect sympathetic, α-adrenergic salivary vasoconstriction. Local stimuli of salivation include the taste, smell and chewing of food in the mouth. Acidic foods, such as lemon juice, are particularly effective and can increase the rate of salivation to about 15 times the unstimulated rate.

Salivation, an unconditioned response or reflex in Pavlov's experiments, can be conditioned by appropriate associations. Although the sight of food makes our mouths water while we are awake, dreaming about food does not stimulate salivation, suggesting that the conditioned salivary response is weaker in man than in dogs.

Most salivary responses are mediated through parasympathetic efferent neurons, from salivatory nuclei in the junction of the medulla and pons. This center is excited by taste and tactile stimuli, and its response can be modulated by facilitatory and inhibitory impulses from the appetite area of the hypothalamus and from the taste-smell areas of the cerebral cortex. Parasympathetic preganglionic fibers in the 7th, 9th and 12th cranial nerves converge to fewer postganglionic fibers in the salivary glands. Parasympathetic stimulation causes copious salivation. Parasympathomimetic drugs increase and atropine decreases salivation. The dry mouth after anesthesia is caused by atropine-like drugs being administered to reduce salivation in order to prevent its aspiration into the lungs.

Sympathetic, β-adrenergic stimulation of salivary glands increases both secretion and blood flow. Thus, with respect to salivary secretion, parasympathetic and sympathetic stimulation may be complementary rather than antagonistic in effect.

Unlike elsewhere in the alimentary tract, the salivary glands are almost totally regulated by their innervation. Hormones have virtually no control over salivary secretion. Although aldosterone can stimulate Na^+-K^+ exchange in salivary ducts, the autonomic nerves control the primary production of acinar fluid and hence the secretory rate of salivary glands, as shown in Figure 8.2. The acini have receptors for acetylcholine as well

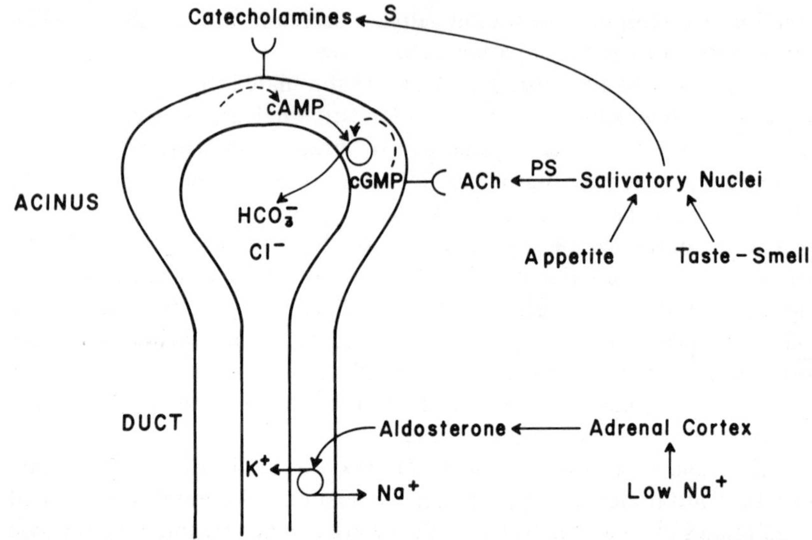

Figure 8.2. Principal regulation of salivary secretion. *ACh*, acetylcholine, released mainly by parasympathetic (*PS*) efferents from salivatory nuclei in medulla-pons; *cGMP*, cyclic guanosine monophosphate; *S*, sympathetic efferents; *cAMP*, cyclic adenosine monophosphate; afferent input from appetite area of hypothalamus and from local taste, smell of food through cerebral cortex.

as the synthetic and degradative enzymes for regulating its effective concentration. Acetylcholine activation of acinar receptors may stimulate intracellular production of a second messenger, cyclic GMP. Similarly, stimulation of β-adrenergic receptors may evoke intracellular synthesis of cyclic AMP. Presumably, the cyclic nucleotides initiate the ultimate stimulation of the anion secretory pumps of the acinar cells.

Summary

Salivary secretion initiates carbohydrate digestion in the mouth and stomach. It contains the required enzyme, α-amylase, as well as the required ions, bicarbonate and chloride, for enzymatic activity. Salivary secretion additionally aids general digestion by providing the lubricant, mucin, for chewing and bolus formation. The bicarbonate of saliva maintains a neutral pH for the teeth which prevents dissolution of tooth enamel by acid-producing bacteria. Formation of saliva is stimulated locally by food in the mouth as well as centrally by salivatory and appetite centers. All stimulation of saliva is finally channeled through the autonomic innervation of salivary glands.

chapter 9

Gastric Secretion

how gastric mucosa produces acid

The stomach is more than a simple storage organ. It secretes copious volumes of a very acidic juice. This acid activates the first proteolytic, digestive enzyme in the gastrointestinal tract, pepsin. The acid-secreting cells also secrete a factor which is essential for the intestinal absorption of vitamin B_{12}, which is essential for life. The protective mechanisms against damage to the lining of the stomach are the gastric mucosal barrier, composed of the mucous coating and membrane integrity of the mucosa, as well as the active ion transporting processes present in the gastric epithelium.

Acid Secretion

A solution of isosmotic hydrochloric acid is elaborated by some one billion oxyntic cells of the human stomach. These oxyntic (or parietal) cells are located along the walls of the oxyntic glands, which are the secretory units of the gastric mucosa (Figure 9.1). Oxyntic cells are structurally specialized for the function of acid secretion as indicated by their possession of numerous mitochondria for energy production and an extensive smooth endoplasmic reticulum for membrane synthesis. They are especially distinguished by their intracellular canaliculi, tortuous invaginations of the luminal membrane across which hydrogen ions are

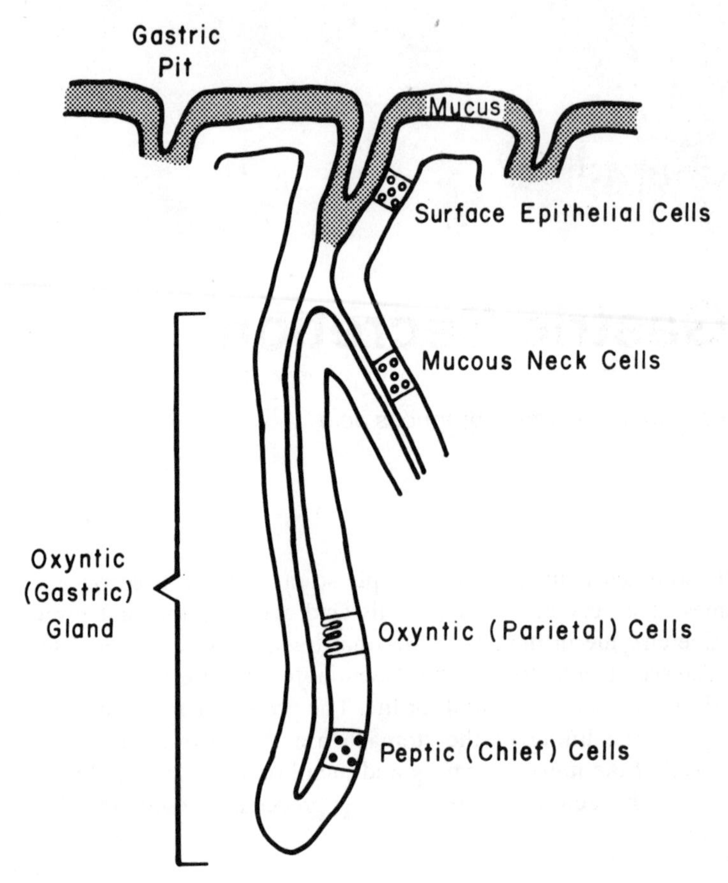

Figure 9.1. The gastric secretory unit. Cells and structures are indicated by both their modern and (classical) names.

transported. The transport of hydrogen ions from a concentration less than 10^{-7} M in plasma to a concentration of about 10^{-1} M in gastric juice is an active mechanism which consumes energy. The essential enzyme for hydrogen ion transport appears to be a $(H^{+}+K^{+})$-ATPase found in oxyntic cell membranes.

The basic process of acid formation in oxyntic cells is the hydrolysis of water into equivalent hydrogen and hydroxyl ions (Figure 9.2). The hydrogen ion is immediately transported across the cell membrane into

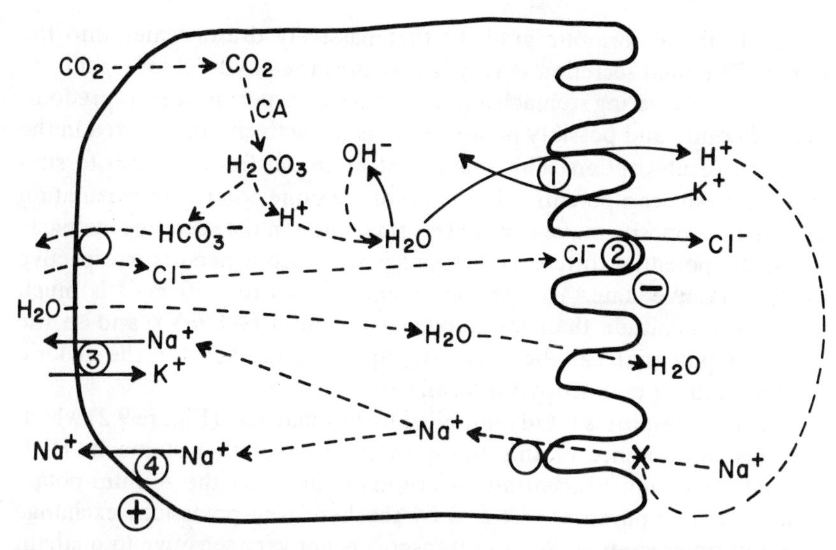

Figure 9.2. Mechanisms of gastric secretion of acid and related processes. ①, $(H^{+}+K^{+})$-ATPase; ②, HCO_3^{-}-ATPase; ③, $(Na^{+}+K^{+})$-ATPase; ④, Mg^{++}-ATPase; ○, facilitated diffusion; *CA*, carbonic anhydrase. Solid arrows indicate active and dashed arrows passive transport.

the gastric lumen by means of the $(H^{+}+K^{+})$-ATPase. The remaining hydroxyl ion needs to be neutralized to avoid toxic alkalinization of the cell. Neutralization occurs through activity of the enzyme, carbonic anhydrase. The gastric mucosa contains large stores of carbonic anhydrase, and high doses of its inhibitor, acetazolamide, inhibit acid secretion. Carbonic anhydrase catalyzes the hydration of carbon dioxide from both plasma and oxyntic cell into carbonic acid. Carbonic acid freely dissociates into hydrogen and bicarbonate ions. The hydrogen ion produced by carbonic anhydrase combines with and neutralizes the hydroxyl ion formed in water hydrolysis. Note that the hydrogen ion secreted is not that produced by carbonic anhydrase activity. The bicarbonate product of carbonic anhydrase activity leaves the oxyntic cell at the basolateral or submucosal border in exchange for plasma chloride. This process tends to increase the pH of gastric venous blood, the so-called "alkaline tide." The chloride is subsequently actively transported across the luminal or mucosal membrane. Active transport of chloride slightly exceeds that of hydrogen, producing a gastric mucosal potential difference with the lumen relatively negative.

The active transport of hydrogen and chloride ions into the gastric lumen forms an osmotic gradient that passively draws water into the lumen. The final secretion is very nearly isosmotic hydrochloric acid.

In the nonsecreting stomach another active transport process is predominant. Sodium and possibly potassium ions are actively transported in the direction of absorption. This movement of cation from mucosa to submucosa produces a potential difference of the same polarity or orientation as does the opposite movement of chloride anion in the secreting stomach. Thus, the potential difference across gastric mucosa persists irrespective of the digestive state. The gastric potential (−40 to −70 mV) is much larger in magnitude than the duodenal potential (−4 mV), and so the electrical potential can be used to provide a marker for the pyloric junction during endoscopy for localization of ulcers.

Sodium transport is totally inhibited by luminal acid (Figure 9.2) which explains the absence of this transport in the secreting stomach. High cellular potassium concentrations are maintained by the sodium-potassium exchange pump at rest and by the hydrogen-potassium exchange pump during secretion. Sodium transport is not very sensitive to ouabain inhibition in the stomach. This cardiac glycoside primarily inhibits active chloride transport through gastric mucosa.

Glucose and fatty acids from gastric mucosal blood are oxidized with the formation of ATP. This ATP fuels at least four distinct gastric mucosal ATPases. Sodium transport is served by Mg^{++}– and (Na^{+}+K^{+})-ATPases, hydrogen transport by a (H^{+}+K^{+})-ATPase and chloride transport by a HCO_3^{-}-ATPase. The HCO_3^{-}-ATPase seems involved in the secretory process because this enzyme, as well as gastric acid secretion, is inhibited by thiocyanate. Much HCO_3^{-}-ATPase, however, is simply mitochondrial.

The hexose monophosphate pentose phosphate (or) shunt is unusually active in gastric mucosa. It is an alternative but incomplete cycle for glucose oxidation. Although it can provide energy for secretion, its function generally seems to involve providing reducing equivalents in the form of NADPH for lipid synthesis. Since lipid synthesis must accompany the canalicular membrane proliferation of acid secretion, the hexose monophosphate shunt is contributory to secretion.

Acid secreted by oxyntic cells is normally superimposed on a basal alkaline secretion probably produced by surface epithelial cells. This alkaline secretion resembles an ultrafiltrate of plasma in lacking protein but including bicarbonate and sodium. Thus, gastric juice is a mixture of two components, a basal alkaline and a stimulated acidic solution. At very low rates of stimulated secretion that occur between meals, the basal alkaline type predominates: the hydrogen concentration is low and the sodium concentration high. At the high rates of stimulated secretion occurring after a meal the acidic type prevails: the hydrogen concentration is high and the sodium ion concentration low. Since chloride ions accompany both acidic and alkaline secretions, chloride concentration of gastric juice is maintained high irrespective of the meal time, or "prandial" state. Similarly, the potassium concentration remains relatively constant, because it also contributes to both types of secretion. These observations are depicted graphically in Figure 9.3.

Appreciable amounts of sodium and potassium are secreted into gastric juice as a result of the basal, constant alkaline secretion. Consequently, when gastric contents are vomited, large amounts of sodium and potassium ions along with even larger amounts of hydrogen ions are lost. Thus, protracted gastric vomiting or continuous medical suction which aspirates gastric juice brings about not only metabolic alkalosis but also hypokalemia (low plasma potassium). Hypokalemia has profound effects on all excitable tissues, especially the conduction system of the heart.

In man, the rate of unstimulated acid secretion is about 2 mEq/hr, with a normal maximum of about 5 mEq/hr. This secretion has a diurnal rhythm with a minimum around 2 a.m. in the morning. Unstimulated acid secretion tends to be lower in gastric ulcer patients and higher in duodenal ulcer patients, but the overlap is considerable and prevents diagnosis on this basis alone. In pernicious anemia the oxyntic cell mass is severely reduced and acid secretion is virtually absent, a condition

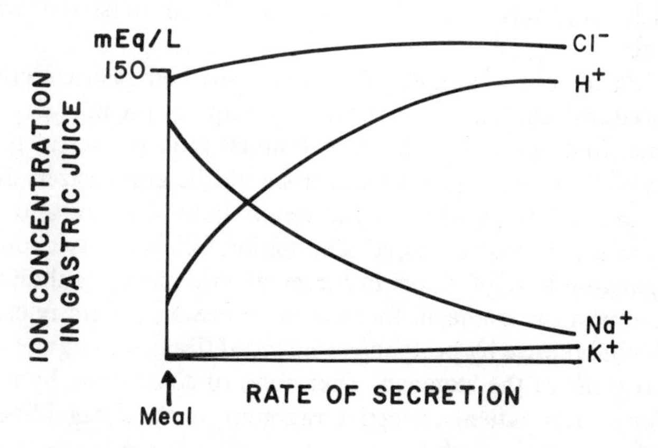

Figure 9.3. Changes in ion concentrations of gastric juice with stimulation of secretion following a meal.

called "achlorhydria." Since the gastric pH does not fall below 6 even with stimulation in patients with pernicious anemia, diagnosis of this condition is suggested from gastric assay.

Besides acid, the oxyntic cells secrete gastric intrinsic factor, a protein that complexes and permits the absorption of vitamin B_{12} in the ileum. Vitamin B_{12} is a maturation factor in erythropoiesis. In the absence of oxyntic cells, the patient develops pernicious anemia and cannot form mature red blood cells.

Acid secretion can be stimulated by three principal "secretagogues": histamine, acetylcholine and gastrin. The actions of these three substances are "synergistic," in that a small dose of one potentiates (more than adds to) the response brought about by a small dose of another. Each has a specific receptor site on the basolateral membrane of the oxyntic cell.

The gastric mucosal histamine receptor is specific to compounds that have an intact imidazole ring structure, the so-called "H-2" requirement. Thus, compounds of this sort can competitively inhibit the action of histamine to stimulate acid secretion. H-2 receptor antagonists like cimetidine are therapeutically useful in controlling the excessive acid secretion of patients with duodenal ulcer.

Another histamine receptor, the "H-1" type, is found in the lungs and elsewhere, but not in oxyntic cells of the gastric mucosa. It can be competitively inhibited by compounds having a positively charged ammonium side chain. H-1 antagonists like tripelennamine are useful in carrying out the gastric function test. In this test, gastric secretory capacity is measured after histamine stimulation. To prevent H-1 side effects such as bronchoconstriction, an H-1 antagonist is administered prior to the histamine.

Since the actions of acetylcholine and gastrin on gastric secretion have been discussed extensively in previous chapters on neural factors and gastrointestinal hormones, the gastric mucosal responses to these secretagogues will not be detailed here. It should be emphasized that acetylcholine can affect gastric acid secretion both directly and indirectly through release of gastrin. Direct stimulation follows acetylcholine release by postganglionic vagal fibers in the myenteric plexus of the fundus and body region of the stomach. Indirect influence on gastrin release follows acetylcholine release by postganglionic vagal fibers serving the antral and pyloric regions of the stomach. Therefore, to control the hypersecretion of duodenal ulcer patients, selective vagotomy of those vagal fibers serving the stomach is performed.

Pepsin Secretion

The principal digestive function of the stomach is the conversion of ingested food into soluble "chyme." This dissolution is accomplished largely by the proteolytic enzyme, pepsin. Pepsin is secreted as a precursor zymogen, pepsinogen, by gastric peptic (or chief) cells. These peptic cells are located on the walls of the oxyntic glands (Figure 9.1). The peptic cells release stored pepsinogen on stimulation by the same secretagogues that stimulate acid; namely, histamine, gastrin, and especially acetylcholine, as well as by secretin. Only acetylcholine and secretin stimulate peptic cells to synthesize and secrete new pepsinogen, however. The inactive pepsinogen released is activated by gastric acid and the pH optimum is 2. In addition, pepsin itself activates pepsinogen. Thus, pepsin activation is an example of positive feedback that leads to rapid and complete conversion of pepsinogen into pepsin (Figure 9.4).

Although pepsinogen is released during feeding and digestion by gastrin and vagal stimulation, activation to pepsin is delayed until the simultaneously stimulated gastric acid is able to overcome food buffers, especially protein. Then pepsin is activated quickly and begins protein digestion. Pepsin digestion of protein is incomplete, cleaving at aromatic amino acid residues to yield various peptides. Pepsin continues to act in the acidic stomach and is only inactivated after passage of chyme into the duodenum where neutralization occurs. Acid activates and base inactivates pepsin (Figure 9.4).

Pepsinogen and pepsin have sufficiently low molecular weights that they can leak through the permeable intestinal mucosa and enter the plasma. Ultimately, they are filtered and excreted into the urine as uropepsinogen. Due to active reabsorption of uropepsinogen protein by pinocytosis in the renal tubules, however, uropepsinogen levels in urine

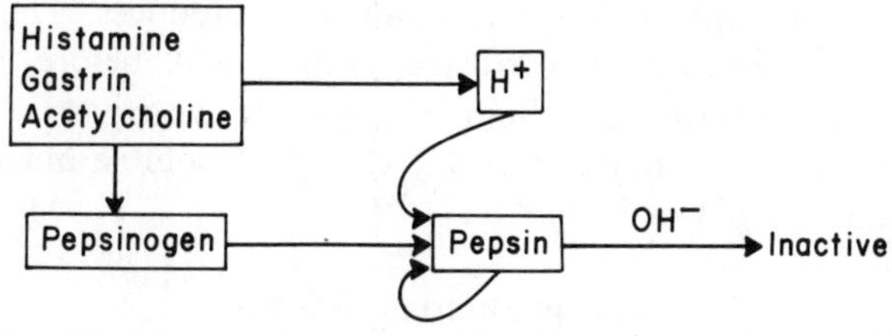

Figure 9.4. Gastric pepsinogen activation as an example of positive feedback.

do not correlate well with levels of gastric pepsin and acid secretion. Hence uropepsinogen concentrations in urine are not useful in the diagnosis of ulcers. Pepsinogens fall into two immunologically distinct groups. Group I pepsinogens are secreted by the peptic and mucous neck cells of the oxyntic gland. Group II pepsinogens derive from pyloric and Brunner's glands as well as from oxyntic glands. Whether one or more of these pepsinogen isoenzymes may serve as possible predictors of ulcers is a subject of interest to gastroenterologists.

Other, more minor digestive enzymes are also secreted by the stomach. The peptic cells secrete gelatinase, which liquifies gelatin. A gastric lipase acts primarily on short-chain length fats. Gastric urease is predominantly if not exclusively bacterial in origin. It converts urea that permeates the gastric mucosa from plasma into ammonia. In the presence of ample acid, the gastric ammonia is trapped as the charged ammonium ion. Normally, the plasma urea level is low, but with renal dysfunction in uremia, however, sufficient ammonia can be produced to neutralize half the gastric acid. Gastric urease activity is abnormal, because gastric acid normally sterilizes chyme, killing the bacteria that produce urease.

Mucus Secretion

Gastric mucus lubricates and protects the gastric mucosa. It is secreted in two forms, a soluble and an insoluble fraction. The soluble fraction is mucoprotein produced by peptic and mucous neck cells (Figure 9.1) in response to vagal stimulation. This fraction mixes with and lubricates gastric chyme during gastric motility. The insoluble mucus is a mucin gel elaborated by the surface epithelium of the gastric fundus and body (Figure 9.1) as well as by the squamous epithelia at the esophageal and pyloric junctions. Besides lubricating and mechanically protecting these tissues against friction with foodstuff, the insoluble mucus provides valuable chemical protection to the underlying cells. Insoluble mucus imbibes the alkaline fluid that the surface epithelium produces. The adhering mucus thereby neutralizes corrosive gastric acid before it reaches the susceptible surface epithelium. On contact with gastric acid, the insoluble mucus precipitates as visible clumps. Insoluble mucus buffers gastric acid in the pH range of 4 to 7.

Gastric Mucosal Barrier

By the term "gastric mucosal barrier" is meant the structural and functional protection of the gastric mucosa against its own secreted acid and pepsin. The structural component includes the physical and chemical barrier of adherent mucus that separates the surface epithelium from gastric acid and the epithelial cell membranes themselves. These components possess a relative impermeability to ions. This low permeability derives both from the small diameter of gastric ion channels in the membrane as well as the tightness of the junctions between cells. Functional protection of the mucosa is afforded by the ion transport mechanisms. By actively transporting hydrogen and chloride out of the mucosa into the lumen, the epithelium is continually removing these ions. Similarly, by continually transporting sodium into the plasma, the gastric mucosa controls the intracellular concentration of this ion.

The gastric mucosal barrier normally protects the epithelium from 0.1 M hydrochloric acid. In animals, it has been possible to bolster the strength of the barrier sufficiently to prevent ulceration even after exposure to 1 M hydrochloric acid or boiling water. Barrier strength is augmented by application or formation of prostaglandins in the gastric mucosa. The mechanism of cytoprotection conferred by prostaglandins is not clear, although prostaglandins stimulate mucus production and sodium transport. Additionally, prostaglandins inhibit acid secretion and increase mucosal blood flow.

Certain ingested and secreted substances in the gastric lumen are capable of breaking down the gastric mucosal barrier. Common barrier breakers include ethanol, salt and sugar at concentrations several-fold hyperosmotic to plasma. Such concentrations of these agents may be achieved by ingestion of martinis, chips and candy. Another potent barrier breaker is aspirin, which is effective at low concentrations. Regurgitated bile salts from the intestinal lumen also break the barrier at low concentrations.

Breakdown of the gastric mucosal barrier by any of the various agents is expressed physiologically by a fall in the luminal concentration of hydrogen ion and a rise in the luminal concentration of sodium ion. These changes may be explained either by an increase in gastric mucosal permeability to ions or by an inhibition of active ion transport through the gastric mucosa (Figure 9.5).

Damaging agents produce intercellular channels that serve to loosen the tight junctions. They also can partially dissolve the lipid bilayer of cell membranes. Both of these actions increase the permeability of the gastric mucosa to ions. As a consequence, hydrogen ions diffuse back from a high concentration in the lumen to low concentrations in surface epithelial cells, where they poison the enzymatic apparatus. As a second

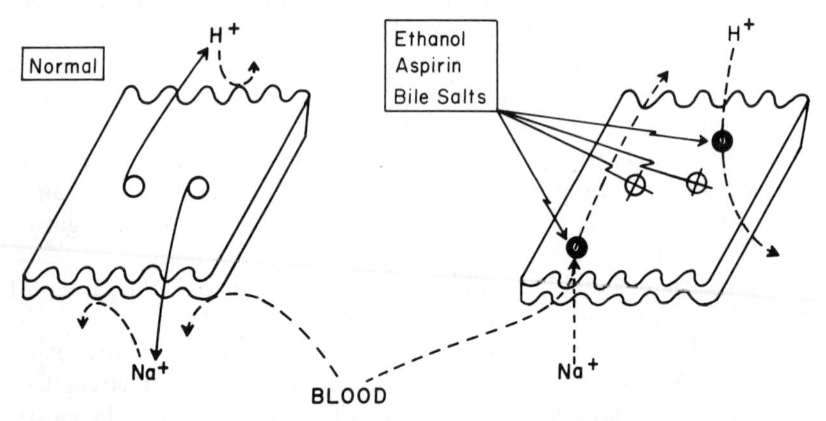

Figure 9.5. Damage to the gastric mucosal barrier by ethanol, aspirin or bile salts. Active transport (*solid arrows*) of hydrogen and sodium ions is inhibited and passive transport (*dashed arrows*) or permeability to these ions is increased.

consequence, sodium ions diffuse from a high concentration in the plasma to low concentrations in the cells and lumen. Increased sodium may also be toxic to gastric cells.

Equally important are the inhibitory effects of damaging agents on active ion transport in gastric mucosa. These agents greatly reduce the active secretion of hydrogen and chloride ions into the lumen and the active absorption of sodium ions from the lumen. Such inhibition also can explain the observations of reduced hydrogen and increased sodium ion concentration after damage. These inhibitory actions on ion transport occur early and may lead to the increased permeability observed later.

The clinical expression of breaking the gastric mucosal barrier by altering active ion transport and permeability of the membranes is commonly observed in our society. There are many individuals who consume strong alcoholic beverages and suffer from vomiting of bile salts and from headaches which are treated with aspirin. Chronic exposure to these barrier breakers leads to inflammation of the gastric mucosa (gastritis), ulcers and even atrophy of the epithelium. Excessive back diffusion of hydrogen ion through the broken barrier may cause local vascular changes which lead to hemorrhage. This form of bleeding is the most common cause of upper gastrointestinal hemorrhage and can be fatal.

Summary

Gastric acid secretion originates with an energy-requiring hydrolysis of water into hydrogen and hydroxyl ions in oxyntic cells. The hydrogen ions as well as chloride ions are actively secreted into the stomach lumen to form isosmotic hydrochloric acid. The hydroxyl ions are converted to bicarbonate ions by carbonic anhydrase and passively enter the gastric venous blood, raising its pH. Acid is essential for activation of the protease pepsin to initiate protein digestion. The oxyntic cells also secrete gastric intrinsic factor, which is necessary for absorption of vitamin B_{12} and normal erythropoiesis. Damage to the gastric mucosa by acid and various ingested and secreted products is prevented by mucosal barriers. Adherent mucus physically and chemically protects the underlying mucosa. Gastric mucosa is normally a tight epithelium, relatively impermeable to hydrogen ions. Finally, active secretion of hydrogen ions removes them from the mucosa where damage would otherwise result.

chapter 10

Pancreatic Secretion

how gastric juice is neutralized and digestion is catalyzed

When gastric acid in chyme reaches the duodenum, the hormone secretin is released from the mucosa into the splanchnic blood. Secretin stimulates the pancreas to produce an alkaline solution that neutralizes the gastric acid in the duodenum. Neutralization is important not only to prevent duodenal ulceration but also to activate digestive enzymes. These enzymes are products of the exocrine pancreas and, in neutralized chyme, they largely complete the breakdown of protein, starch and fat.

Bicarbonate Secretion

The pancreatic secretion of a fluid containing a high concentration of bicarbonate occurs in a system of acini and ducts that resemble the salivon of the salivary glands both structurally and functionally. The blind sacs, or acini, of the "pancreaton" drain into a branched ductular system that converges finally into the main pancreatic duct. The main pancreatic duct in man may enter the proximal duodenum separately or join with the bile duct to empty into a common duct before entering the intestine.

The critical site for pancreatic secretion of bicarbonate appears to be the duct cells, especially those near the acini. The first step in production of bicarbonate is the active transport of hydrogen ions out of the duct cell into the interstitial fluid on the blood side (Figure 10.1). This acidification

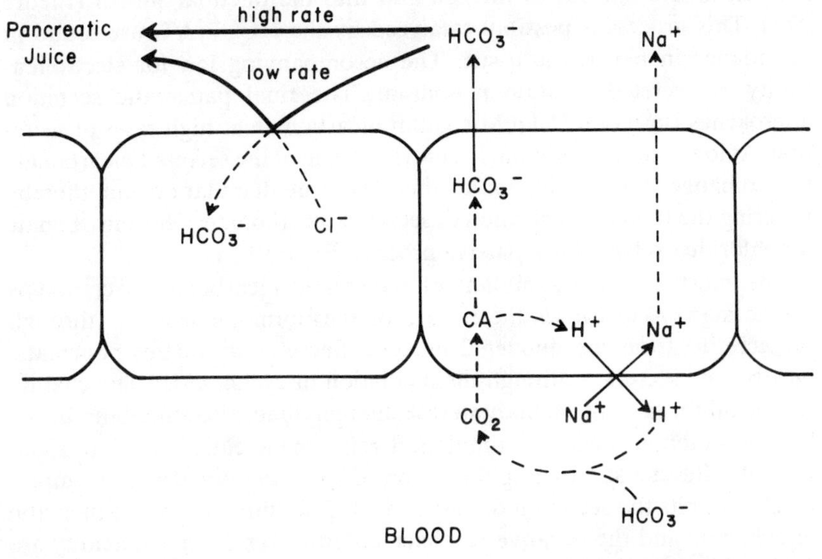

Figure 10.1. Pancreatic secretion of bicarbonate. *CA*, carbonic anhydrase. *Solid arrows* indicate active and *dashed arrows* passive transport. *Heavy arrows* are flows.

of pancreatic venous blood, the so-called "acid tide," neutralizes the alkaline tide of gastric venous blood, thereby restoring acid-base balance to the splanchnic vasculature. The hydrogen ion is pumped out of the duct cell in exchange for a sodium ion that is pumped into the cell. This active exchange for sodium accounts for the sensitivity of pancreatic bicarbonate secretion to agents like ouabain and ethacrynic acid that inhibit epithelial transport of sodium.

The hydrogen ions in pancreatic plasma convert bicarbonate ions to carbon dioxide, which is lipid-soluble and permeable in the membranes of the ducts (Figure 10.1). After diffusing into the cell, the carbon dioxide is converted back to bicarbonate ions by means of the enzyme, carbonic anhydrase. This latter step accounts for the sensitivity of pancreatic bicarbonate secretion to acetazolamide, a characteristic inhibitor of carbonic anhydrase. The other product of carbonic anhydrase activity is a hydrogen ion that can subsequently be pumped out of the cell and recycled.

The final step in pancreatic bicarbonate secretion is the active transport of bicarbonate ion out of the cell and into the ductular lumen (Figure 10.1). This process is possibly catalyzed by the HCO_3^--ATPase found in abundance in pancreatic tissue. The accompanying ion for electroneutrality of secreted solution is sodium. The final pancreatic secretion approaches isosmotic 150 mM sodium bicarbonate at high rates of secretion. At low rates of secretion, in contrast, some of the secreted bicarbonate ion exchanges for chloride ion further down the ductular system, thereby lowering the final concentration of secreted bicarbonate. This bicarbonate for chloride exchange is a passive process (Figure 10.1).

The most effective stimulant of pancreatic bicarbonate secretion is gastric acid. Acid causes the release of the hormone secretin (through prosecretin) from the duodenal mucosa. Secretin stimulates pancreatic bicarbonate secretion through the mediation of cyclic AMP. The characteristic inhibitor of phosphodiesterase, theophylline, also stimulates bicarbonate secretion. Secretin stimulation raises pancreatic secretion about 150-fold. Besides stimulating formation of base, secretin also noncompetitively inhibits the secretion of gastric acid. The stimulation of pancreatic bicarbonate and the negative feedback control over duodenal acidity are depicted in Figure 10.2.

The vagus contains cholinergic fibers that can weakly stimulate pancreatic bicarbonate secretion. Digestion products of protein and fat in chyme also weakly stimulate.

Adrenergic innervation is inhibitory to pancreatic bicarbonate secretion. Although both α- and β-adrenergic neurons are inhibitory, the α type alone has been implicated in the etiology of duodenal ulcer. Cigarette smokers have a higher incidence of duodenal ulcer than do nonsmokers. This difference could result from either increased acid secretion or decreased bicarbonate secretion in smokers. In concentrations equivalent to smoking 4 cigarettes/hr, nicotine has no effect on gastric acid secretion. These same concentrations of nicotine, however, reduce pancreatic bicarbonate secretion 3-fold. Nicotine inhibition of pancreatic bicarbonate secretion is mediated by sympathetic ganglionic discharge to the pancreas. Norepinephrine mimics nicotine inhibition and an α-adrenergic inhibitor, phenoxybenzamine, blocks nicotine inhibition.

Enzyme Secretion

The secretion of pancreatic bicarbonate solution makes possible the secretion and activity of several pancreatic enzymes. The ionic secretion

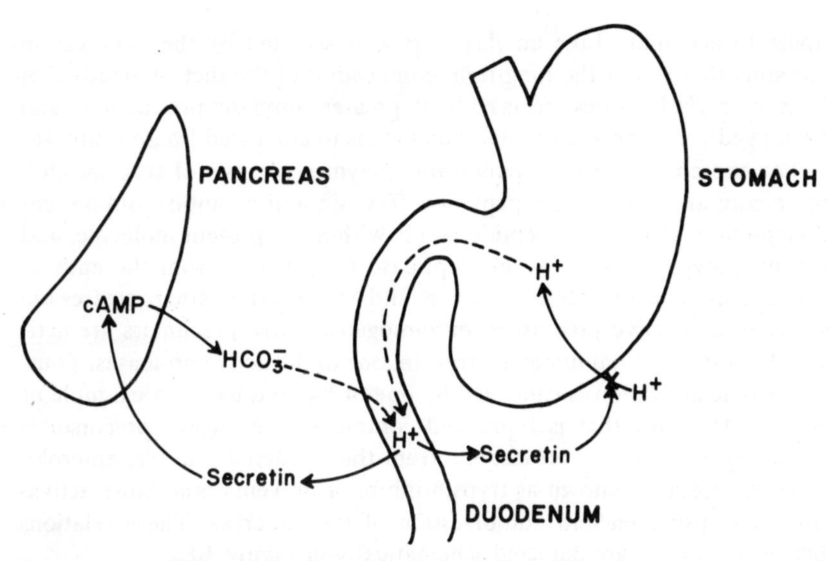

Figure 10.2. Stimulation of pancreatic bicarbonate and the neutralization of gastric acid. *cAMP*, cyclic AMP.

of the pancreas is isosmotic to plasma and volumetrically important. The 1½ liters of stimulated pancreatic juice per day washes out the enzymes that are secreted into the acini without large accompanying osmotic equivalents of water. Without secretin stimulation, these enzymes would remain useless in the lumina of the acini and ducts of the pancreas. After being washed out into the intestinal lumen, the enzymes require a neutral pH for their digestive activity. The accompanying bicarbonate solution provides this essential condition.

Pancreatic enzymes comprise three classes, based upon the foodstuff digested. The amylytic class acts on starch and has just one member, amylase. It hydrolyzes starch to maltose, maltotriose and dextrins. Pancreatic amylase is secreted in the active state and completes the action of salivary amylase. Pancreatic amylase activity increases throughout infancy until full pancreatic function is attained around age 3.

The lipolytic class of pancreatic enzymes consists of a lipase and phospholipases. Pancreatic lipase is secreted in the active state and hydrolyzes triglycerides to free fatty acids, monoglyceride and glycerol. Its action is facilitated by bile salt emulsification of fat droplets into much smaller particles, micelles, that have a much greater total surface area for lipase to act upon. Like amylase, lipase is secreted by the pancreas in amounts that reflect the long-term compostion of the diet. A steady diet high in carbohydrates brings about greater amylase production, and prolonged consumption of a fatty diet leads to enhanced lipase synthesis.

The proteolytic class of pancreatic enzymes consists of trypsin, chymotrypsin and carboxypolypeptidase. Trypsin and chymotrypsin are endopeptidases that cleave peptide bonds within the protein molecule, and carboxypolypeptidase is an exopeptidase that cleaves from the ends of the protein molecule. Unlike amylase and lipase, pancreatic proteases are secreted as inactive precursors, or zymogens. These precursors are activated in the intestinal lumen by trypsin, one of the active proteases. Thus, proteolytic activation is catalyzed by one of its products—an example of positive feedback that is rapid and complete. The trypsin precursor is originally activated by an enzyme from the duodenal mucosa, enterokinase. A substance known as trypsin inhibitor prevents premature activation of trypsinogen and autodigestion of the pancreas. These relations between enzymes are depicted schematically in Figure 10.3.

Pancreatic enzyme activity accounts for most of the digestion of carbohydrates, fats and proteins into their respective products: mono- and disaccharides, fatty acids and monoglycerides, dipeptides and amino acids. The more minor enzymes of pancreatic juice digest phospholipids and nucleotides. Pancreatic enzymatic activity is fully sufficient for complete digestion. Salivary amylase and gastric pepsin are not essential.

Normally the digestive enzymes of the pancreas act in the lumen of the upper gut; under abnormal conditions, however, these enzymes may attack the tissue of the pancreas itself, producing a catastrophic and life-threatening inflammatory state called pancreatitis. As the pancreatic substance is autolysed, local hemorrhage and great pain result and a fatal shock state may ensue. Pancreatitis is associated in many cases with either long-standing consumption of excessive amounts of alcoholic beverages or with preexisting inflammation of the gallbladder.

The primary stimuli of pancreatic enzyme secretion are acetylcholine and cholecystokinin. Acetylcholine is released by the vagal nerve endings in the pancreas. The vagus is stimulated during the cephalic phase of digestion by the perception of food. Cholecystokinin (or pancreozymin, as it used to be called) is released from APUD cells in the mucosa by digestion products in the duodenal lumen. Essential amino acids and fatty acids are particularly potent releasers of cholecystokinin. The action of cholecystokinin is mediated by cyclic GMP, which releases intracellular calcium ions. Calcium ions activate a Ca^{++}-ATPase in zymogens that

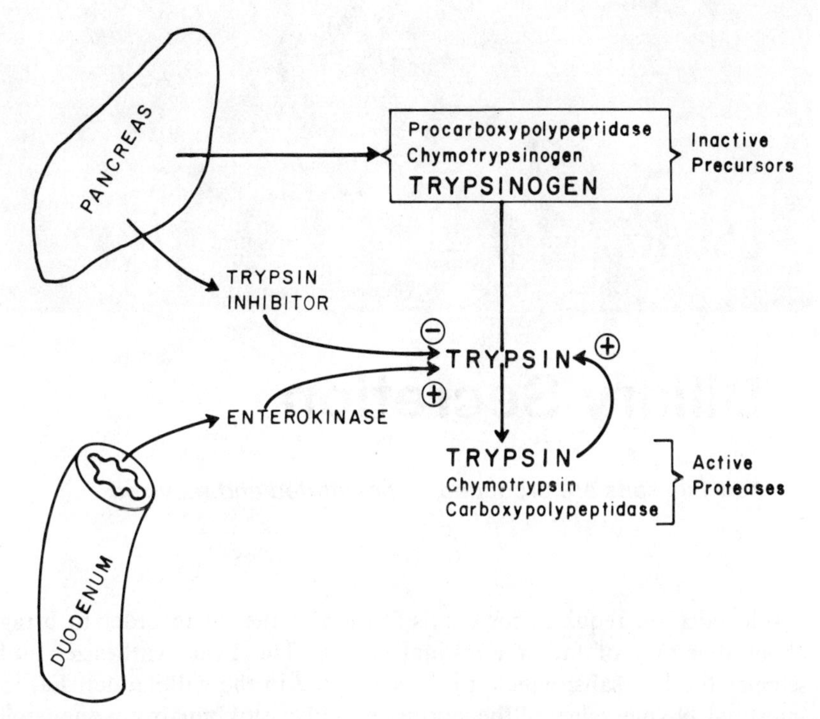

Figure 10.3. Activation of pancreatic proteolytic enzymes by trypsin. *Positive signs* refer to positive feedback control and *negative sign* to inhibition.

permits fusion of zymogen and cell membranes and brings about exocytosis of the enzyme package.

Summary

Pancreatic secretion has two principal components that interrelate. A voluminous alkaline secretion by the pancreas is stimulated by the presence of potentially ulcerating gastric acid in the duodenum. The response is hormonal, mediated by secretin, and homeostatic—the duodenal lumen is neutralized. In the neutralized lumen the enzymatic secretion by the pancreas is active. This enzymatic secretory product is stimulated both neurally, by the vagus, and hormonally, by cholecystokinin. The enzymes secreted by the pancreas are washed into the duodenum by the bicarbonate secretion of the pancreas. Pancreatic enzymes are essential for the complete digestion of carbohydrate, fat and protein.

chapter 11

Biliary Secretion

how bile salts are produced, concentrated and recycled

Bile salts are required for emulsification of the fat in order to bring about digestion of fat in intestinal chyme. The liver synthesizes and secretes the bile salts, which are concentrated in the gallbladder. Fat in intestinal chyme releases the hormone cholecystokinin from duodenal mucosa, and cholecystokinin stimulates contraction of the gallbladder and emptying of the bile salt solution into the intestinal lumen. After emulsifying fat in the duodenum and jejunum, bile salts are actively absorbed in the ileum and returned to the liver for re-secretion.

Hepatic Secretion

Bile acids are synthesized from cholesterol in the liver. The liver synthesizes only the primary bile acids, cholic and chenodeoxycholic. Secondary bile acids, deoxycholic and lithocholic, are products of bacterial metabolism of primary bile acids in the intestine.

Prior to secretion, both primary and secondary bile acids are conjugated with the amino acids, taurine or glycine. Conjugation makes the bile acids more water-soluble by adding a negative charge. The conjugated bile salts are both more soluble in the watery chyme of the intestine and more readily absorbed by the ileum. (Conjugated bile salts are actually less readily absorbed by the duodenum and jejunum, which keeps them in the lumen of that portion of the small intestine where digestion is occurring. Consequently, deconjugation by bacteria in blind loop syndrome causes lowering of bile salt concentration in the lumen and leads to steatorrhea.)

Since conjugated bile salts precipitate out of solution when the pH falls below 4, it is important that the accompanying ionic solution of biliary secretion be neutral or alkaline. A fluid resembling plasma ultrafiltrate and containing bicarbonate is secreted by bile ducts. The primary stimulus for this biliary bicarbonate secretion is the hormone secretin, which simultaneously stimulates pancreatic bicarbonate secretion. In these ways, conjugated bile salts are maintained soluble both in the liver and in the intestine.

Both conjugated bile salts and sodium ions are actively transported across canalicular membranes of the bile capillaries that converge into the hepatic duct (Figure 11.1). Bile salts are secreted by a transport maximum (T_m)-limited mechanism, but the substrate concentration of bile salt in hepatic plasma is the physiological limitation on normal secretion. Sodium ions are transported actively by a (Na^++K^+)-ATPase. Of the choleretics or substances that stimulate bile formation, the most important are the plasma concentration of bile salts, the hormones secretin and cholecystokinin, and vagal stimuli. Flow of bile in the canaliculus is countercurrent to blood flow in the sinusoids (Figure 11.1). Thus, hepatocytes high on the biliary tree remove bile salts from relatively low plasma concentrations and those low on the biliary tree remove bile salts from relatively high plasma concentrations at rates near the T_m.

A parallel, T_m-limited secretion of another organic substance occurs in hepatocytes. Bilirubin, the metabolic end-product of hemoglobin catabolism by reticuloendothelial cells, is actively secreted into the bile canaliculi (Figure 11.1). Like bile salts, bilirubin is first extracted from sinusoid blood and then conjugated prior to secretion. The extraction process involves the transfer of bilirubin from its plasma binding protein, albumen, to the cytoplasmic binding proteins Y and Z. Bilirubin is conjugated with glucuronate within the hepatocyte.

In the intestine, bilirubin is metabolized by bacteria to urobilinogen, which, when oxidized in air, forms urobilin. Urobilin is one of the compounds that makes the stool brown and urine yellow.

Two other organic substances of importance, cholesterol and lecithin, are also secreted into bile canaliculi. The details of their secretion, however, are not known.

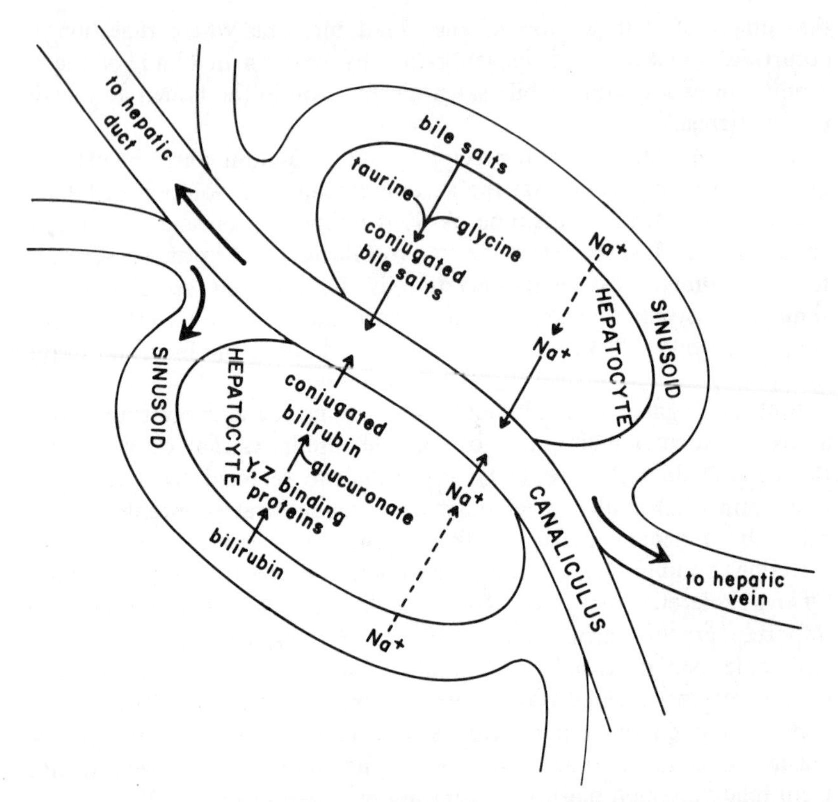

Figure 11.1. Secretory unit of liver. Active transport is indicated by *solid arrows,* passive transport by *dashed arrows* and flow by *heavy arrows.*

Gallbladder Concentration

In man but not all animals, biliary secretion is stored and concentrated in a diverticulum of the hepatic duct called the gallbladder. The gallbladder concentrates bile salts and the other organic substances by the active absorption of electrolytes and water. Absorption of ions and fluid follows the active transport of sodium ions from the lumen of the human gallbladder (Figure 11.2). Chloride ions closely follow the transported sodium ions, and in some animals transport of the two is coupled by a common carrier. Bicarbonate ions also follow transported sodium, and the concentration of bicarbonate in bile is maintained close to its plasma value, about 25 mM. As in all epithelia, ions are transported from the lumen of the gallbladder into the lateral, intercellular spaces and only

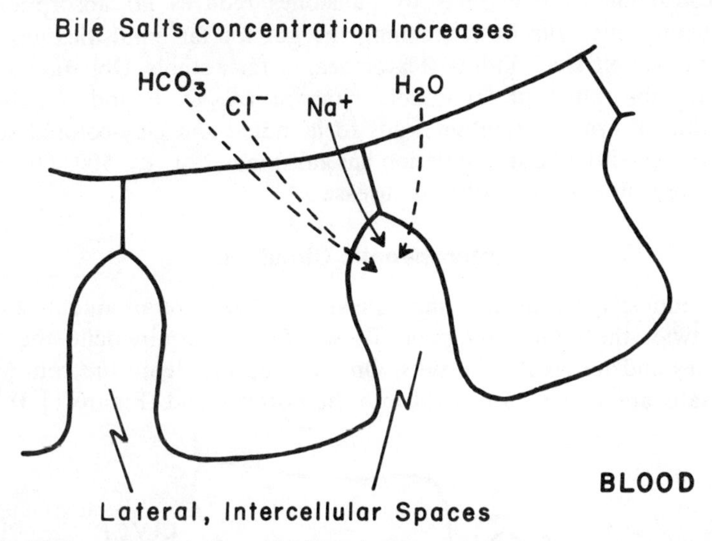

Figure 11.2. Concentration of bile salts through absorption of electrolytes and water by gallbladder epithelium. *Solid arrows* indicate active and *dashed arrows* passive transport.

subsequently into the blood. A standing osmotic gradient for the passive absorption of water may be created by such intercellular ion concentration. Absorption of fluid and electrolyte concentrates bile salts, bilirubin, cholesterol and lecithin about 20-fold in the gallbladder. The concentrated bile is approximately isosmotic to plasma, nevertheless, because the organic constituents of the micelles are osmotically almost inactive.

Due to the considerable concentration of organic substances brought about by the gallbladder, precipitates called "gallstones" sometimes form in gallbladder bile. Gallstones are usually composed of cholesterol, principally. They generally form when the biliary cholesterol concentration is abnormally high or when primary biliary bile salt concentration is abnormally low. Bile salts incorporate cholesterol (and lecithin) into micelles and normally prevent cholesterol precipitation. Cholesterol saturation of bile increases during the night, because no meals stimulate the gallbladder to empty. Biliary cholesterol saturation can be reduced by administration of chenodeoxycholate or of another bile salt, ursodeoxycholate. Large scale testing of these agents is underway to see whether gallstones can be dissolved. Administration of bile salts induces both an increase in bile salt secretion and a decrease in cholesterol synthesis and secretion.

Obstruction of bile ducts by gallstones reduces fat absorption by preventing bile salts from reaching the gut for fat emulsification. The result is the physical finding steatorrhea, or fatty stools. Obstruction also impairs absorption of fat-soluble vitamins (A, D, E and K) and of bilirubin. Bilirubin retention leads to jaundice and clay-colored stools. About one-tenth of our population has gallstones; last year 500,000 people underwent surgery for gallstone disease.

Enterohepatic Circulation

To emulsify the fat in a single meal would require an amount of bile salts twice the total body pool. To satisfy this requirement, the body recycles and reuses its bile salts. On reaching the ileum the conjugated bile salts are actively absorbed into the portal blood (Figure 11.3). Bile

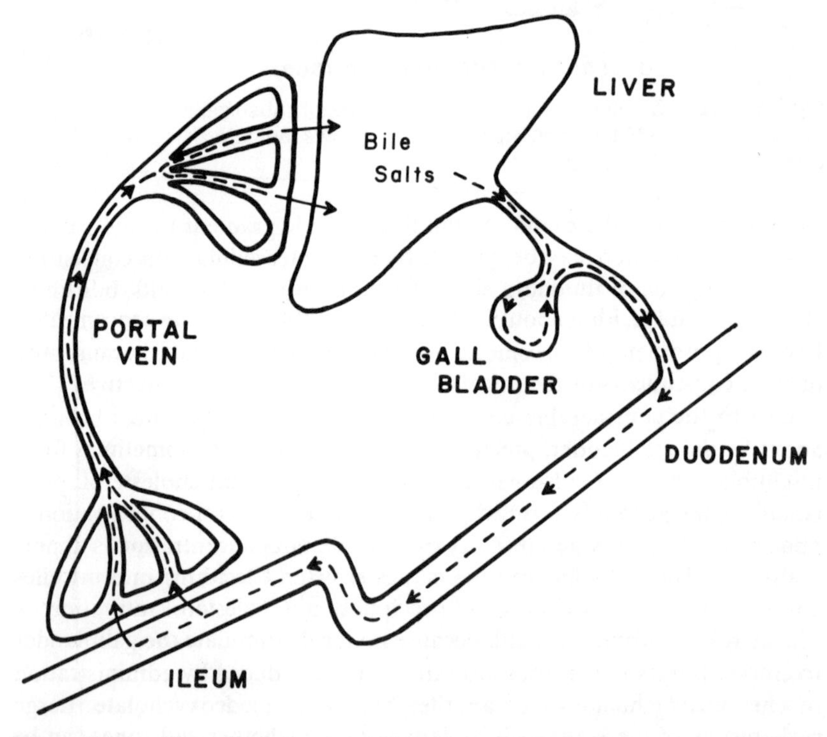

Figure 11.3. Enterohepatic circulation of conjugated primary and secondary bile salts. *Solid arrows* indicate active and *dashed arrows* passive transport.

salts that have been deconjugated by intestinal bacteria, being lipid-soluble, are passively absorbed throughout the small intestine. Both types of bile salts are returned via the portal vein to the liver. The hepatocytes respond to the increased plasma concentration of bile salts with an increased rate of uptake. Re-secretion of bile salts into the bile canaliculi ensues. On the re-secretion, both primary and secondary bile salts are added to bile. This recycling process for bile salts and a few other substances is called the "enterohepatic circulation." Other agents of medical importance which are handled in the enterohepatic circulation include indomethacin and phenolsulfonphthalein.

A small fraction of the total bile salt pool is lost because some bile salts are not reabsorbed in the ileum. This leak of less than 5% of the bile salt pool per circulation is replaced by equivalent bile salt synthesis in the liver. The unreabsorbed bile salts are deconjugated and metabolized by bacteria in the colon. When the loss is excessive, the bile salt concentration can rise sufficiently high to inhibit sodium ion transport and water absorption by the colon, leading to a watery diarrhea. The total concentration of primary and secondary bile salts in the intestinal lumen averages 15 mM.

In man, the bile salt pool is equally divided among cholate, chenodeoxycholate and deoxycholate. Lithocholate is sulfated in the human liver and not recycled. There appear to be defects in hepatic handling of lithocholate in patients with gallstones.

To facilitate fat digestion after a meal, pancreatic lipase as well as biliary bile salts are recycled. Pancreatic enzymes are actively absorbed in the ileum and transferred by way of the portal blood to the pancreas for re-secretion. Thus, the enteropancreatic circulation of enzymes is analogous to the enterohepatic circulation of bile salts.

As indicated previously, the presence of excessive amounts of fat in the stools, steatorrhea, is abnormal. Beside the loss of calories and essential fatty acids, fat-soluble vitamins are also wasted. Furthermore, unabsorbed fat in the gut forms soaps with Ca^{++} and Mg^{++}, thereby reducing the absorption of these minerals. The presence of excess solute in the gut attracts water osmotically and leads to diarrhea with loss of water-soluble solutes as well.

Steatorrhea will result from interruption of the complex enterohepatic circulation which may occur in many ways. Intentional interruption of the enterohepatic circulation has been utilized to lower intestinal absorption of cholesterol, a component of fat which is thought to be etiological in atherosclerosis. Drugs are employed which cause bile salts to precipitate in the intestinal lumen, thereby reducing their reabsorption in the ileum. The bile salt pool shrinks and less cholesterol is absorbed.

Interruption of the enterohepatic circulation may also occur with inflammatory disease of the mucosa of the ileum (Crohn's disease), as a result of surgical bypass of the ileum (treatment of morbid obesity), with extensive inflammation of the hepatocytes (hepatitis) or with toxic damage to those cells as a result of exposure to industrial chemicals such as vinyl chloride or carbon tetrachloride.

Summary

Cholic and chenodeoxycholic acids, the primary bile acids, are synthesized from cholesterol in the liver. They are conjugated to taurine or glycine and secreted into the bile canaliculi. Conjugation and bicarbonate secretion in the bile ducts maintain the solubility of the bile salts. Before entering the intestine, bile salts are concentrated in the gallbladder as a result of active sodium absorption. In the intestine the bile salts serve their principal function of emulsifying fat. While in the intestine, the bile salts may be metabolized by resident bacteria. Bacteria convert the primary bile salts to secondary types, deoxycholate and lithocholate. They also can deconjugate bile salts. Deconjugated bile salts are passively absorbed, but the conjugated variety is specifically absorbed by an active mechanism in the ileum. Both conjugated and deconjugated bile acids are returned by portal blood to the liver for re-secretion, a process known as the enterohepatic circulation.

chapter 12

Water and Electrolyte Absorption

how intestinal mucosa absorbs secreted and ingested fluids

The intestinal mucosa absorbs virtually all of the large volume of fluids and dissolved electrolytes that are secreted by salivary glands, the stomach, the pancreas, the liver and by the intestine itself, as well as absorbing most of the water and electrolyte that is ingested. This formidable task, to be referred to as "hydroelectrolytic absorption," is accomplished simply and efficiently by the process of active sodium transport, which passively draws along anions and water. When this sodium absorptive mechanism is overcome by secretory processes stimulated by intestinal microorganisms or distension, diarrhea results.

Intestinal Secretion

The intestinal mucosa releases three kinds of secretion into the gut lumen: mucus, sloughed mucosa and fluid. Mucus is secreted by goblet cells throughout the intestine in response to stimulation by the vagus and by local distension or irritation. Mucous secretion lubricates the epithelial surface and protects it from mechanical damage by solid foodstuff. Mucus secreted by Brunner's glands in the proximal duodenum may additionally protect the mucosa from chemical damage due to the influx of acid from the stomach.

As discussed previously in the chapter on mucosal metabolism, intestinal mucosa normally sloughs or desquamates into the gut lumen at high rates. The sloughed mucosa contains active disaccharidases and dipeptidases that contribute to intestinal digestion and absorption of carbohydrates and protein. Sloughed mucosa does not, however, contribute to the hydroelectrolytic absorption for several reasons. First, hydroelectrolytic absorption involves transfer from lumen to blood across an intact epithelium, from which sloughed mucosal material detaches. Second, the enzyme energizing hydroelectrolytic absorption, $(Na^{+}+K^{+})$-ATPase, is inactive after sloughing due to cell lysis and cytoplasmic loss that accompany desquamation. Cellular fragments from sloughing may, in fact, somewhat retard hydroelectrolytic absorption by raising the colloid osmotic pressure of the gut lumen. Products of sloughed disaccharidase activity also can increase luminal osmotic pressure and hinder absorption of water and electrolytes.

Intestinal crypt cells normally secrete an alkaline fluid composed largely of isosmotic sodium chloride and bicarbonate. This basal secretion of the intestinal crypt cells is ordinarily absorbed in its entirety by the intestinal villus cells. Such a recycling of fluid may serve to solubilize chyme at the digestive surface.

Under the influence of toxins produced by certain microorganisms, intestinal secretion may be stimulated to levels far greater than the absorptive capacity of the intestine. Cholera exotoxin, for example, stimulates active chloride and bicarbonate secretion to such a degree that the normal active sodium absorption (Figure 12.1*A*) is overcome (Figure 12.1*B*). With the reversal in osmotic gradient produced by the active chloride and bicarbonate secretion, water streams into the intestinal lumen and a watery diarrhea ensues. Not only the massive diarrheas of asiatic cholera but also some of the more common bacterial diarrheas, such as *Escherichia coli* and *Salmonella*, also appear to work by this mechanism. These diarrheagenic strains produce toxins that stimulate formation of cyclic AMP in intestinal mucosa. Naturally occurring substances in the intestinal mucosa, such as prostaglandins and vasoactive intestinal peptide (VIP), also stimulate synthesis of cyclic AMP and intestinal secretion. VIP is produced in tremendous amounts by an endocrine tumor of the pancreas and prostaglandins may be secreted by medullary tumors of the thyroid. These diseases are characterized by a massive, watery diarrhea.

Choleraic secretions of the intestine may be reversed by stimulating the largely intact absorptive mechanisms. The glucocorticoid drug, methyl prednisolone, overcomes choleraic secretion by stimulating intestinal

A. Normal

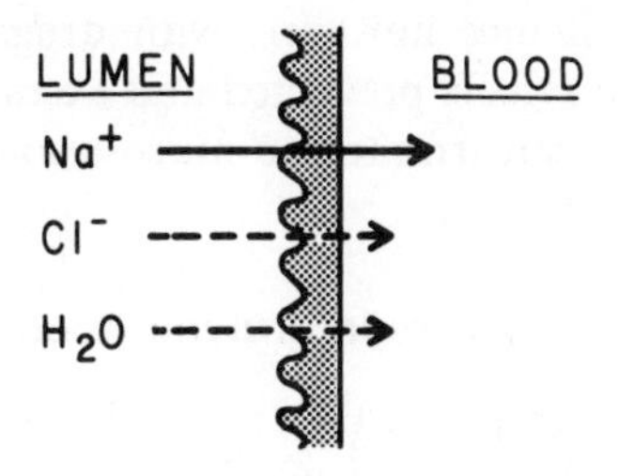

B. Choleraic

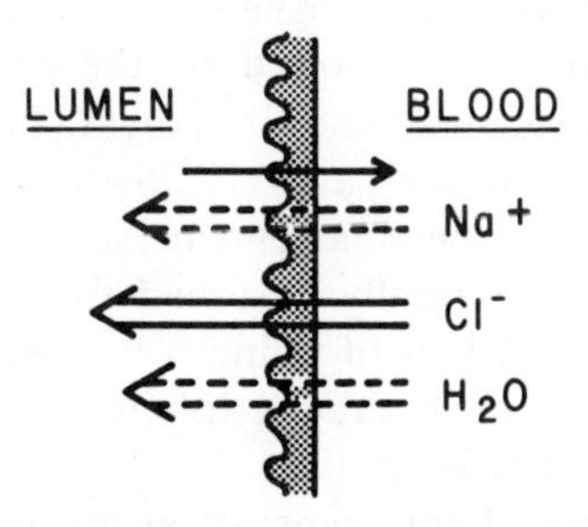

Figure 12.1. Movements of sodium, chloride and water through normal (*A*) and choleraic (*B*) intestinal mucosa. *Solid arrows* indicate active and *dashed arrows* passive transport.

$(Na^+ + K^+)$-ATPase to absorb more sodium. Since most sodium absorption in the intestine is coupled to the absorption of glucose or amino acids, infusion of a solution with glucose through a nasogastric tube into the intestinal lumen stimulates the sodium pump to overcome the opposing chloride pump.

With obstruction of a portion of the intestine, absorption but not secretion is inhibited. Overdistension and reflex contraction inhibit blood flow to the normally hypoxic villus cells that absorb electrolyte and water. The secreting crypt cells are not affected and continue to add volume at the site of obstruction, further distending the segment. As much as a third of the extracellular fluid can accumulate in this way at an obstructed intestinal segment. Without surgical intervention and aspiration of this fluid, the vicious cycle repeats itself and kills the patient.

Small intestinal secretion is stimulated by the duodenal hormones, secretin and cholecystokinin, and inhibited by catecholamines. An important stimulant of large intestinal secretion is the concentration of bile salts that leak from the ileum. Normally 95% of the bile salts are actively absorbed in the ileum; however, with drugs such as cholestyramine, absorption of bile salts is prevented and excessive amounts pass into the colon where they can irritate the mucosa and inhibit hydroelectrolytic absorption.

Hydroelectrolytic Absorption

Active sodium transport is the principal driving force for the absorption of the main electrolytes and water from the lumen of the small and large intestine. Sodium is actively transported against both the electrical and the chemical gradients that normally prevail across the intestinal mucosa. Transport of sodium is followed by the passive transport of chloride flowing down its electrochemical gradient (Figure 12.1*A*). Water follows both sodium and chloride passively down the osmotic gradient which is created. As a result of these movements, some 8½ liters of largely isosmotic sodium chloride is normally absorbed from the human small intestine each day. Another ½ liter of saline is normally absorbed from the large intestine. Without such hydroelectrolytic absorption, severe diarrhea and dehydration would rapidly ensue.

Several features of this general mechanism of intestinal absorption require further explanation. The sodium absorptive mechanism is not one but a family of active pumps. The simple, independent pumping of sodium ions by a (Na^++K^+)-ATPase localized in the basolateral membranes of intestinal villus cells accounts for only about 20% of the total sodium absorbed. The largest fraction of sodium absorbed, some 80% of the total, is coupled to one of a variety of other transported solutes. Perhaps the most important of the cotransported solutes are the monosaccharides, especially glucose. Because sodium and glucose attach to a common carrier at the luminal membrane of the brush border, subsequent entry of these substances into the cell is greatly facilitated. Sodium is then transported out of the cell by an active mechanism and glucose subsequently diffuses out of the cell through the basolateral membrane into the interstitium and blood. This coupled sodium and sugar absorption will be discussed in greater detail in the following chapter on carbohydrate absorption.

Sodium absorption is also coupled to the absorption of amino acids in the intestine. Other common carriers for these species facilitate their entry at the brush border and their subsequent absorption. The details of this process will be discussed in the chapter on protein absorption.

Finally, a significant fraction of sodium is directly coupled to chloride in intestinal absorption. In this case, sodium and chloride enter the cell by means of a common carrier.

As for all instances of coupled transport, the addition of one cotransported species greatly enhances the absorption of the other. For the example cited previously, the addition of glucose to the choleraic lumen stimulates sodium absorption from the intestine. Coupling of sodium absorption to that of these other common nutrients in the diet insures the rapid and complete absorption of all solutes and water.

Absorption of monosaccharides and amino acids is virtually complete on passage through the duodenum. Absorption of electrolytes and water is also rapid, but complete absorption requires the entire length of the intestine. The reasons for this are:

1. Whereas nutrients entering the intestine derive exclusively from the ingested diet, most electrolytes and water in the intestine derive from secretions rather than food. Each day the gut absorbs the equivalent of 9 liters of isotonic saline of which about 1 liter is derived from ingested food and fluids. The remaining 8 liters/day is secreted into the gastrointestinal tract as saliva (1 liter/day, mostly hyposmotic $NaHCO_3$), gastric acid (2 liters/day, mostly isosmotic HCl), pancreatic and biliary juice (2 liters/day, mostly isosmotic $NaHCO_3$) and intestinal juice (3 liters/day, mostly isosmotic NaCl). Since intestinal secretion enters all along the intestine, complete absorption of electrolytes and water generally requires the entire intestinal length.

2. Different transport mechanisms in various portions of the intestine allow for the selective entry of electrolytes. In the ileum and colon, chloride is actively absorbed in exchange for actively secreted bicarbonate, the latter serving to neutralize acids produced by resident bacteria in those regions of the gut. In the colon, active sodium absorption generates a large electrical potential difference with the lumen negative. This electrical gradient causes passive secretion of potassium ions. Just as in the salivary ducts and distal nephron, this absorption of sodium and secretion of potassium in the colon can be stimulated by the adrenal hormone, aldosterone. Due to the specialized secretion of bicarbonate and potassium by the colon, excessive loss of intestinal fluid in diarrhea brings about metabolic acidosis and hypokalemia.

The flows of the principal electrolytes into and out of the gastrointestinal tract are portrayed in Figure 12.2. Note that of the total 9 liters/day of water and electrolyte normally entering the intestine, all but about 90 ml, or 1%, is subsequently absorbed. The proportion of fluid reabsorbed

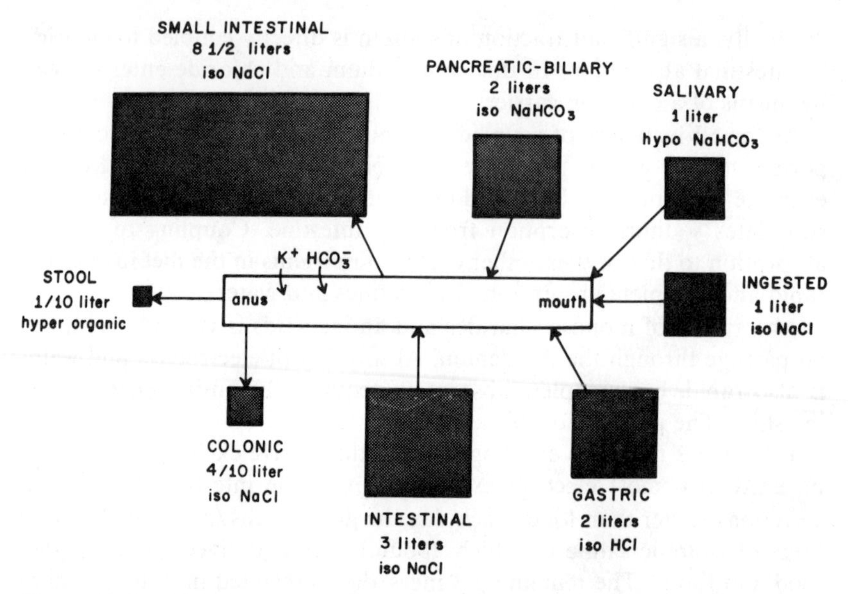

Figure 12.2. Movements of water and electrolyte into and out of the intestine. *iso*, isosmotic; *hypo*, hyposmotic; *hyper*, hyperosmotic.

by the gut is, therefore, very comparable to that found in the kidney. Intestinal absorption of electrolytes is even more efficient than that of water: only 1 mEq each of sodium and chloride and about 5 mEq of potassium ions are excreted each day in the stool. Organic ions produced by bacteria, however, make the soluble stool somewhat hyperosmotic.

The colon, like the stomach, is also a storage organ with a "tight" epithelium and small pores (4 Å) which restrict passive diffusion of ions and osmosis of water. This tightness allows for some charge separation between actively absorbed sodium and passively accompanying chloride, giving rise to a large potential difference. (The rate of passive absorption of chloride is over twice the rate of active absorption of chloride in exchange for bicarbonate.) A second consequence of the tight colonic epithelium is the hyperosmolality of the stool. Water does not immediately enter the colon as bacteria produce osmotic equivalents. The result is conservation of water when bowel movements are normal.

In contrast to the colon and stomach, the small intestine has a "leaky" epithelium having loose tight junctions between cells and large pores (7 to 15 Å). These conditions permit very close coupling of the actively absorbed sodium and the passively accompanying chloride, yielding only a small potential difference. The loose epithelium also permits ready osmotic equilibration between the plasma and small intestinal contents, which are normally very isosmotic to plasma. When the small intestine does occasionally become hyperosmotic, water enters rapidly to distend the gut. The overdistension leads to nausea and the so-called "dumping" syndrome. The intestinal contents can be made hyperosmotic surgically as a result of a gastro-jejunal fistula that bypasses the pyloric sphincter and allows hyperosmotic gastric contents to enter the especially sensitive jejunum too rapidly. In sensitive individuals during the course of normal digestion of a high carbohydrate meal, the small intestine can become hyperosmotic due to the relatively faster rate of carbohydrate digestion into many osmotically active sugars as compared with their slower rate of absorption.

Osmotic equilibration in the small intestine is also the basis of cathartic or laxative action. Poorly absorbed salts such as magnesium sulfate retain their osmotic equivalent of water in the intestine, thereby increasing the volume of unabsorbed fluid in the gut, which in turn, causes distension and diarrhea.

Absorption of the major electrolytes and water proceeds largely unregulated by neural and hormonal influences. Catecholamines do somewhat stimulate the one small fraction of sodium that is coupled to chloride during absorption. This α-adrenergic stimulation of sodium chloride absorption, however, is balanced by a reduced bicarbonate absorption. Principal regulation of hydroelectrolytic absorption in the small intestine, therefore, resides in the substrate concentrations of intestinal electrolytes and cotransported nutrients. At physiological concentrations aldosterone does not influence sodium absorption and potassium secretion, and any action of aldosterone is limited to the end organs of the alimentary canal, the salivary glands and the colon.

Overall, the function of the intestine is to absorb all electrolytes and water. Regulation of electrolyte and water balance in the body is accomplished mainly by the kidney. To insure the absorption of nearly all electrolytes and water, the intestine has a large reserve capacity. This can, however, be exceeded during secretory diarrhea and even by ingestion of large volumes of hypotonic fluid in a short time.

Most of the small intestine can be resected in cases of gross obesity before absorption of nutrients is appreciably limited. The ileum alone suffices for normal absorption, since the ileum is the most specialized absorptive segment. (The ileum is the sole site of conjugated bile salt and vitamin B_{12} absorption.) After proximal intestinal resection, the ileum responds with epithelial hypertrophy which again provides greater absorptive mucosa and added reserve capacity.

Calcium and Iron Absorption

In contrast to the almost complete and mostly unregulated absorption of the common monovalent electrolytes, the divalent cations calcium and iron are absorbed incompletely and in a regulated fashion in the intestine. Their rates of absorption depend critically on the body's balance of these ions. When more calcium or iron is utilized or excreted than is made available or ingested, hormonal feedback mechanisms stimulate enhanced absorption. When a positive balance of calcium or iron prevails, on the other hand, the same negative feedback mechanisms reduce intestinal absorption of these ions.

Calcium is absorbed in the duodenum as a divalent cation. Most consumed calcium is not ionic and not soluble at neutral pH. Gastric acid solubilizes the calcium salts and permits their absorption from the duodenum. Duodenal mucosa contains a calcium-binding protein and a Ca^{++}-ATPase that are responsible for the active transport of calcium across the basolateral membrane. Calcium attaches first to the calcium-binding carrier protein at the brush border membrane. This facilitated transport process is saturable but only at concentrations of calcium greater than would ordinarily be found in the intestinal lumen. Calcium detaches from the carrier and enters the cytoplasm of the cell. The concentration of calcium in the cell is maintained very low by the continual mitochondrial uptake and active transport of calcium out of the cell through the basolateral membrane. This latter process is energized by the Ca^{++}-ATPase. Note that the active extrusion of calcium out the basal side promotes the concentration gradient for facilitated diffusion of calcium from the lumen.

The facilitated entry and the active absorption of calcium are closely regulated by a metabolite of vitamin D in accord with overall calcium balance. The metabolite is produced under the influence of parathyroid hormone, whose own production is subject to direct negative feedback by the plasma concentration of calcium. When plasma calcium falls below its normal concentration (5 mM or 10 mg/100 ml total; half ionic), the parathyroid glands secrete increased parathyroid hormone to mobilize calcium from body stores and to stimulate absorption of calcium through vitamin D. Parathyroid hormone catalyzes the formation of the active metabolite of vitamin D (1,25-dihydroxycholecalciferol) in the kidney. The metabolite of vitamin D acts directly on intestinal mucosa to increase absorption of calcium. When the concentration of calcium in plasma is excessive, all these steps are reversed. Less parathyroid hormone is produced, less vitamin D metabolite is formed and less calcium is absorbed. Regulation of calcium absorption is depicted in Figure 12.3.

Iron is also absorbed as a divalent cation in the duodenum. Most ingested iron consists of insoluble ferric compounds. Gastric acid, intrinsic factor and/or mucosubstance solubilize and reduce the ferric to the ferrous ion, thereby preparing iron for absorption from the duodenum. Ferrous ions are actively transported from the lumen into the mucosal cell, from which the ions may be actively transported from the cell to the plasma. In the plasma ferrous ions attach to transferrins, their plasma carrier proteins. This is the predominant pathway for absorption in states of iron depletion, such as hemorrhage, where iron is needed for increased erythropoiesis and increased iron absorption contributes to the production of needed hemoglobin. Since the hypoxia accompanying blood loss releases the hormone erythropoietin from the kidney, enhanced iron absorption may possibly be a response to erythropoietin.

Although a negative iron balance is the most important regulator and stimulant of iron absorption, specific intraluminal factors may also enhance or reduce this absorption. Iron complexed to heme is more readily absorbed than is inorganic iron, perhaps due to the greater lipid solubility and permeability of heme iron. In the intestinal cell, heme oxygenase releases inorganic iron. Certain ingested compounds also appear to complex iron in the stomach and promote iron absorption in the duodenum. These substances include ascorbate, alcohol and fructose. Other ingested substances such as phosphate and oxalate form insoluble iron salts and reduce iron absorption. The regulation of intestinal iron absorption is portrayed in Figure 12.4.

In states of positive iron balance, absorbed iron is stored in mucosal cells immediately following initial uptake. The ferrous ions are oxidized to ferric, and granules called "ferritin" form. This iron storage in the intestine is the mucosal block of iron absorption. The iron storage pathway responds to the iron stores of the body. Excess or "messenger" iron appears to influence intestinal cells that are forming in the crypts, so that after they become villus cells they form ferritin. Ultimately, the villus cells slough into the intestinal lumen and the stored iron in them is eliminated. In cases of pathological iron overload, or siderosis, the obligatory iron loss due to normal desquamation is supplemented by a compensatory iron loss due to greatly increased iron stores in the sloughed cells. When the iron overload is very severe, cells called "siderophages" collect excess peritoneal iron by phagocytosis, migrate through the intestinal wall and excrete iron into the lumen.

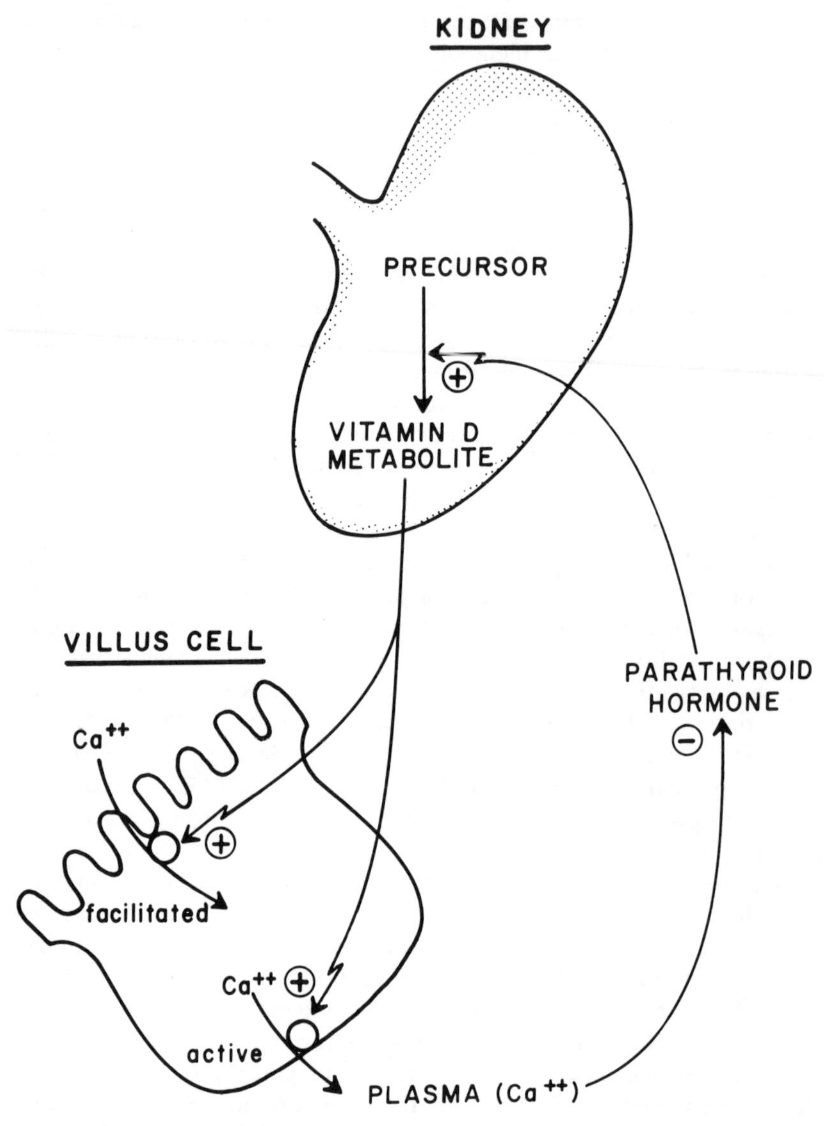

Figure 12.3. Regulation of intestinal calcium absorption. Precursor = 25-hydroxycholecalciferol; vitamin D metabolite = 1,25-dihydroxycholecalciferol. ⊕, positive feedback; ⊖, negative feedback.

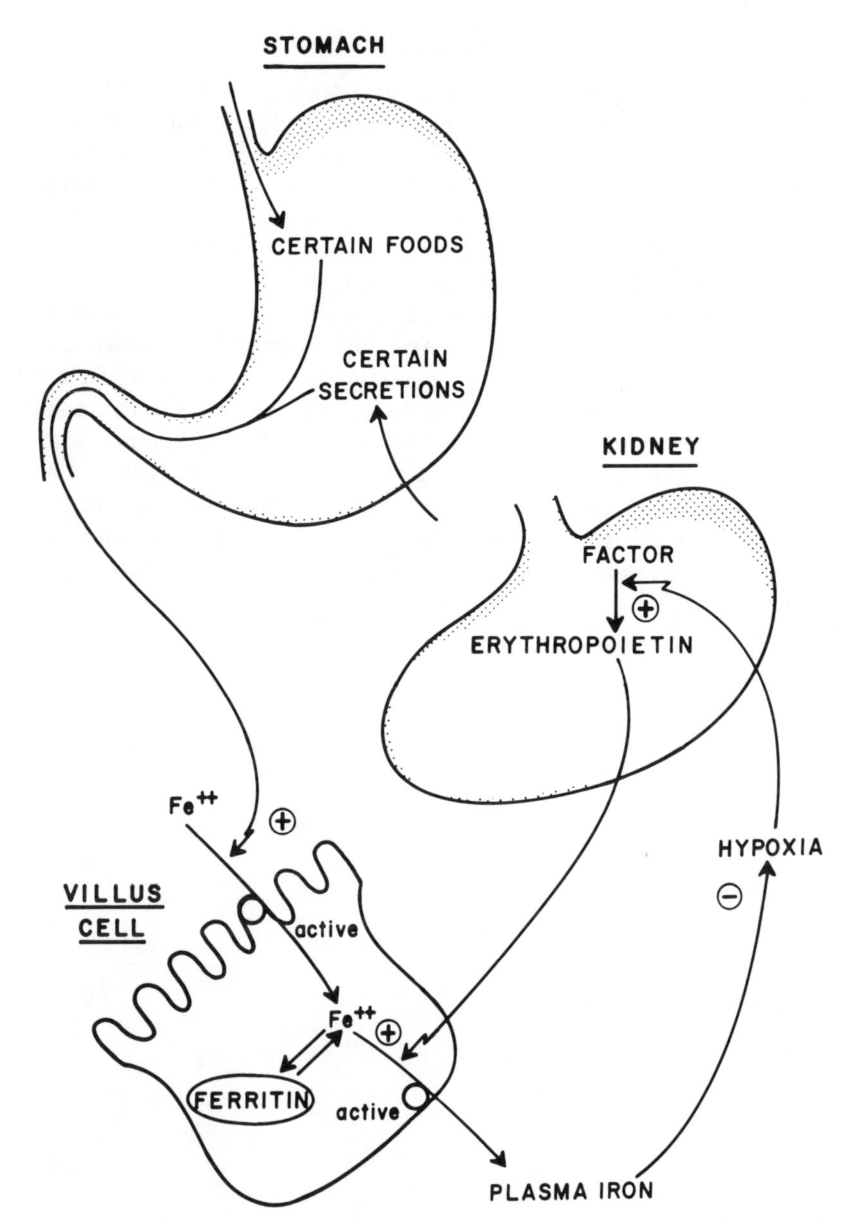

Figure 12.4. Regulation of intestinal iron absorption. ⊕, positive feedback; ⊖, negative feedback.

Summary

Intestinal absorption of water and electrolytes is the normal net balance of many absorptive and secretory processes operating simultaneously. The predominant process is the active absorption of sodium that carries along chloride and water. This active sodium absorption is stimulated by digested nutrients such as monosaccharides and amino acids. Normally, almost all of the large volumes of secreted and smaller volumes of ingested fluid are completely absorbed by this mechanism. There is little hormonal or neural control of this process in the intestine. (Potassium and bicarbonate are secreted in the large intestine, and diarrhea leads to depletion of potassium and bicarbonate.) In contrast to the mostly complete and unregulated absorption of the major electrolytes and water, calcium and iron are absorbed incompletely under hormonal and intraluminal control. Calcium absorption is regulated by parathyroid hormone activity through a metabolite of vitamin D and iron absorption may be controlled by erythropoietin. Constituents of gastric juice and ingested substances solubilize these divalent cations in preparation for their absorption in the duodenum.

chapter 13

Carbohydrate Absorption

how intestinal mucosa absorbs sugars

Tushar K. Chowdhury

One of the most important functions of the gastrointestinal tract is absorption of the digested food, and the principal site of absorption is the small intestine. Efficient absorption of all major types of food depends upon the following factors:

1. The degree of digestion of the food prior to absorption. Most types of food remain unabsorbed or are at best very poorly absorbed in the undigested or intact form. In the pathological state called lactase deficiency, milk sugar is not digested and is not absorbed.

2. The extent of the absorptive surface area. The smaller the surface area, the less the rate of absorption in a given time. The small intestine is so favorable a site for absorption because of the enormous surface it presents. The surface of the small intestine possesses special architectural features, namely, the folds of Kerckring, villi and microvilli, which amplify its absorptive surface area, compared to that of a smoothy lined tube. The folds of Kerckring alone multiply the effective surface area by a factor of 3; the villi of the small intestine further increase the absorptive surface 10-fold; and the microvilli on the epithelial cell membrane increase the effective surface area by an additional factor of 20. Altogether, the

luminal surface of the small intestine presents an effective surface area equivalent to the surface of a tennis court. Any pathological or corrective surgical condition which greatly reduces this absorptive surface area also interferes with absorption. In the disease sprue, for example, where the microvilli and even the villi are absent, intestinal absorption is greatly reduced.

3. The effectiveness of the transport mechanisms. In sprue the loss of the absorptive membranes also means the loss of active transport mechanisms, which further reduces absorption.

4. The time the food spends in the small intestine. If the transit of food is rapidly accelerated through the small intestine, there will be insufficient time for effective digestion and absorption. Rapid transit times are observed after the surgical procedure of vagotomy.

5. A form of motor activity in the intestine called "rhythmic segmentation." This particular type of motility mixes the contents of the small intestine enhancing exposure to the absorptive surface of the small intestine.

Methods for Studying Intestinal Absorption

There are several *in vivo* and *in vitro* techniques for studying intestinal absorption. Some of the useful techniques to study carbohydrate absorption include:

1. Tied Loop Technique *in Situ*

This method is useful in animal experiments. The abdomen is opened under anesthesia and one or more segments of the small intestine are isolated by ligatures. These loops are then filled with a known volume of test solution containing a known concentration of a specific sugar. Following a prescribed period of time the solution is removed, its volume determined and measurement is made of the concentration of the sugar. The difference in the amounts of sugar (volume × concentration) between measurements is expressed as the rate of absorption of the sugar during that period of time.

2. *In Vitro* Perfusion Technique

This method is also used in animal systems. A portion of excised small intestine is isolated, cannulated at both ends and bathed in a chamber having adequate oxygenation. The lumen of the intestinal segment is perfused with a solution containing a known amount of a particular sugar.

Analyses of the sugar concentrations and the volumes of luminal and bathing solutions at various periods can give an estimate of the magnitude of transport of sugars from the lumen to the serosal bathing solution. A modification of this procedure is the so-called "everted sac" technique in which the luminal surface of the intestine faces outward. The major advantage of this technique lies in the fact that the small volume of the serosal solution makes it easier to detect absorption of small amounts of sugar.

3. Intubation Technique

In patients, a double-lumen polyvinyl or polyethylene tube can be passed into a desired segment of the small intestine. The test solution is perfused through one tube into the proximal end of the intestinal segment, whereas the second tube is used to aspirate the distal end of the segment. In order to determine accurately the net movement of water and solutes from the perfusate, it is necessary to dissolve a nonabsorbable marker (*e.g.*, polyethylene glycol) in the test solution.

Absorptive Processes

Absorption of digested food from the lumen of the intestine consists of the transfer of various digestive products across the intestinal epithelium into the blood and the lymph. This chapter deals with the absorption of carbohydrates from the human small intestine. The specific questions posed here are: In what chemical forms are the carbohydrates absorbed? What is the daily load of absorption? What is the site of absorption? What are the mechanisms of absorption?

The average daily intake of carbohydrates of an adult is about 250 g. These carbohydrates are primarily starch (potatoes, rice, spaghetti, pastry and bread), sucrose (table sugar), lactose (milk sugar), fructose (fruit sugar) and glucose (honey). The amount of carbohydrate secreted into the lumen of the gastrointestinal tract is negligible compared to that taken in as food. Therefore, the total daily load of carbohydrate absorbed by the small intestine is equal to that which is eaten.

When the liquified and partly digested food (chyme) is emptied into the duodenum, the dietary carbohydrates are subjected to the enzymes pancreatic amylase and some intestinal amylase. The resultant digestion products are primarily the disaccharides maltose and isomaltose, maltotriose, some α-dextrins and small amounts of the monosaccharides glucose, galactose and fructose. The intestinal mucosa is relatively impermeable to the disaccharides, that is, only monosaccharides can be absorbed across the mucosa into the portal blood. It has been established that the disaccharides are digested into monosaccharides and absorbed simultaneously at the brush border or microvilli of the mucosal columnar epithelial cells. A class of enzymes, the disaccharidases (maltase, lactase, sucrase) and α-dextrinase, are located in the brush border.

There are also small amounts of disaccharidases which are found in the luminal contents of the small intestine. These enzymes are derived from desquamated cells coming from the tips of the intestinal villi. The surface epithelial cells move from the depths of crypts to the tips of the villi where they are shed into the lumen. In man it takes 4 to 5 days for cells of the duodenal and jejunal crypts to complete this journey. Paneth cells and enterochromaffin cells turn over much more slowly.

Digestion by the brush border enzymes (Figure 13.1) yields as major products glucose, galactose and fructose, and these products ultimately must be transported across the intestinal mucosa into the general circulation. Since glucose presents by far the largest sugar load to the intestine (80% of the total, Figure 13.1), it is appropriate to discuss glucose absorption in detail. With the use of both *in vitro* and *in vivo* techniques it has been established that only small quantities of glucose diffuse passively across the intestinal epithelial cells down a concentration gradient; the major portion of glucose transport across the intestinal mucosa occurs against a concentration gradient. This uphill transport of glucose across the intestinal epithelium is dependent upon the metabolic energy of the tissue. Glucose absorption is drastically reduced by metabolic inhibitors, by hypoxia and by decreased temperatures. In the adult human

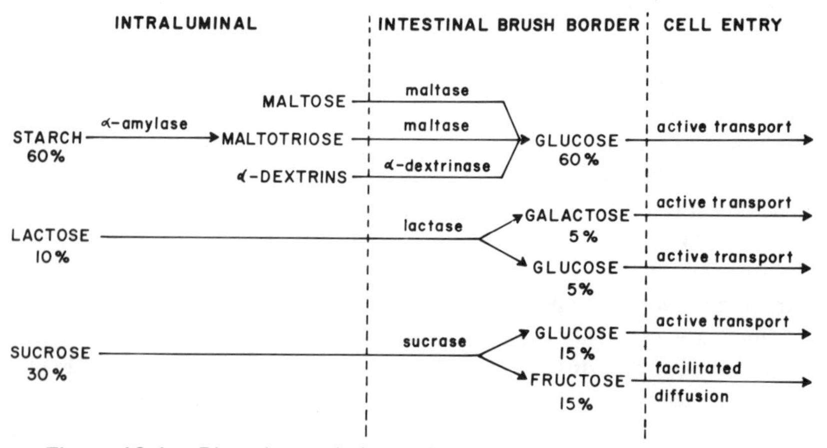

Figure 13.1. Digestive and absorptive activities at the brush border.

as well as in animals the metabolic energy-supplying process involves oxidative phosphorylation and therefore a deficiency in the oxygen supply to the intestinal tissue adversely affects the intestinal glucose absorption. In certain newborn animals, however, glycolytic processes provide the energy for the sugar absorption and consequently in these newborns, intestinal glucose absorption is not dependent upon the tissue oxygen content. In human adults, it has been observed that glucose absorption reaches a maximum rate when the brush border glucose concentration reaches 140 mM. Because of the preceding considerations, glucose absorption is thought to occur through some active transport mechanism.

One of the most interesting facets about active glucose absorption is its dependence upon the presence of Na^+ ions in the intestinal lumen. Furthermore, chemical interference with Na^+ transport with agents such as cardiac glycosides results in a corresponding reduction in glucose absorption. It has also been demonstrated that Na^+ transport itself is augmented by glucose and by its nonmetabolized analogs. It has been postulated, therefore, that glucose transport and Na^+ transport are coupled, that is, they are co-transported by a common carrier molecule (Figure 13.2). The sugar molecule and Na^+ are bound to the common mobile carrier at the luminal border of the microvilli. The carrier transports both to the other surface of the membrane where glucose and Na^+ are released. In terms of this model each carrier molecule (a specific component of the microvillous membrane) has two binding sites, one for

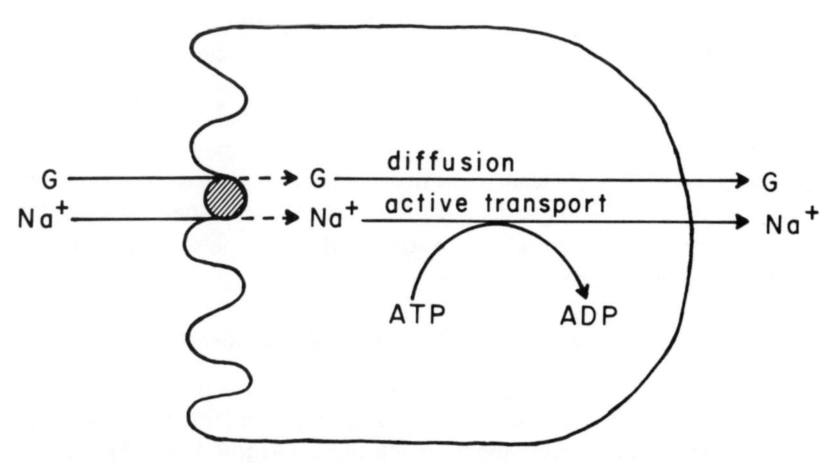

Figure 13.2. Carrier-mediated Na^+-coupled transport of glucose (*G*).

sugar and the other for Na^+. When the Na^+ site is occupied, the carrier molecule exhibits increased affinity for the sugar molecule and then the complex is ready to travel across the membrane. Once the carrier-complex has translocated to the intracellular surface, both Na^+ and the sugar are released. On the intracellular side of membrane where the concentration of K^+ is high, K^+ occupies the Na^+-binding site. In this configuration the carrier molecule has very low affinity for the sugar and the K^+-carrier complex moves back to the luminal surface where K^+ is released and another pair of Na^+ and glucose attach to the carrier molecule to begin their journey. The questions which come up at this juncture are: How does the sugar molecule then reach the general circulation and how is the metabolic energy coupled to the sugar absorption process itself? On the basis of the results of various *in vitro* studies, currently accepted answers to these questions are presented schematically in Figure 13.2. The intracellular Na^+ concentration is maintained at a low level by the ATP-dependent active transport of Na^+ across the serosal membrane. The energy-coupled Na^+ extrusion maintains a concentration gradient of Na^+ favoring further cotransport of glucose and Na^+ into the cell. In this way the intracellular glucose concentration rises and ultimately the sugar diffuses out of the cell through the serosal side into the general circulation. Thus, metabolic energy is indirectly coupled to intestinal sugar absorption.

With regard to the specific site of absorption of glucose some autoradiographic studies suggest that the absorption of this sugar occurs mostly in the upper third of the intestine. Galactose is also absorbed across the same region and primarily by active transport. Some experimental evidence suggests that galactose transport involves the same carrier system as glucose. In the presence of sufficient Na^+ a high concentration of galactose in the intestinal lumen competitively inhibits the active transport of glucose and *vice versa.* Both glucose and galactose transport are inhibited by such metabolic poisons as 2,4-dinitrophenol, phlorizin and ouabain. Fructose absorption across the intestinal mucosa does not involve metabolic energy from the tissue even though a specific carrier system is involved in its transport. The downhill transport of fructose from the lumen of the small intestine occurs through the so-called passive facilitated transport process.

Pathophysiology of Carbohydrate Absorption

Any genetic deficiency of a disaccharidase can lead to malabsorption of monosaccharides because of inadequate digestion. In disaccharidase deficiency the nonabsorbed sugars reach the distal intestine where there is bacterial breakdown of these undigested disaccharides into lactic and other acids and they increase the osmotic attraction of water into the lumen. This situation leads to diarrhea. Lactase deficiency is not uncommon and exclusion of milk from the diet of these individuals may completely relieve diarrhea and other symptoms. Disaccharidase deficiencies are routinely associated with the disorders characterized by atrophy or inflammatory desquamation of the small intestinal epithelium. Such enzyme deficiencies have been noted in patients with enteritis and sprue.

Intestinal malabsorption of carbohydrates is often assessed grossly by measuring the rate of urinary excretion of a nonmetabolized sugar, *d*-xylose. This sugar is normally absorbed through the mediation of the same carriers which are responsible for glucose and galactose absorption. Following almost complete absorption in the jejunum, *d*-xylose is cleared from the blood by the kidneys and excreted in the urine. Therefore, in a patient with normal kidneys, any decrease in the amount of *d*-xylose appearing in the urine suggests malabsorption of glucose and galactose in the small intestine, *e.g.*, sprue.

Carbohydrates are normally absorbed in the duodenum and the proximal half of the jejunum. Disaccharidase activities in the ileum are normally much lower than those in the duodenum and jejunum. However, it has been observed that, in patients with jejunal resection, the ileum adapts itself with heightened disaccharidase activities.

Summary

Efficient absorption of carbohydrates is dependent upon (1) the degree of digestion of the food prior to absorption, (2) the extent of the absorptive surface area, (3) the effectiveness of the transport mechanisms, (4) the time food remains in the small intestine and (5) the mixing action of the gut itself. The average daily intake of carbohydrates in adults is about 250 g in the form of starch, sucrose, lactose, fructose and glucose. The intestinal mucosa is relatively impermeable to the disaccharide form of carbohydrates. Disaccharides are digested into monosaccharides and absorbed simultaneously at the membranes of the brush border of microvilli. The enzymes disaccharidases (maltase, lactase and sucroase) and α-dextrinase are located in the brush border. Glucose makes up about 80% of the breakdown products of carbohydrates and is absorbed by a sodium-dependent active transport mechanism. Galactose shares the same carriers as glucose does. Both glucose and galactose absorption are inhibited by metabolic poisons and hypoxia. Fructose is absorbed by a passive facilitated transport process. A genetic deficiency of a disaccharidase can lead to malabsorption of the corresponding sugars which in turn leads to diarrhea. Sugars are normally absorbed in the duodenum and the proximal half of the jejunum.

chapter 14

Protein Absorption

how intestinal mucosa absorbs peptides and amino acids from protein sources

The small intestine normally absorbs all of the ingested protein as well as secreted and sloughed protein. Some enzymatic protein and trace amounts of ingested protein may be absorbed intact by pinocytosis. The principal route of protein absorption, however, involves digestion into small peptides and amino acids. These small peptides and amino acids are then actively transported across intestinal mucosa into the portal blood. The transport mechanisms for amino acids are rather specific, and clinical disorders involving malabsorption of even a single amino acid have been described.

Sources of Protein

To achieve nitrogen balance the 70 kg adult must eat 50 g/day of protein. The growing child requires proportionately more protein relative to body weight. Except for tiny amounts of allergenic protein that is absorbed intact by pinocytosis, all ingested protein is digested prior to absorption in the adult and child.

In the newborn, γ-globulins from colostrum and milk can be absorbed intact, but they are subsequently digested within intestinal mucosal cells. Although γ-globulin absorption confers passive immunity to newborn of lower animals, the gut of the human infant is not permeable to detectable amounts of colostral γ-globulins. Human infants acquire their passive immunity earlier during the fetal stage, in which maternal γ-globulins are transferred across the placenta to the fetus.

γ-Globulins constitute more than half of the protein found in colostrum. They derive unaltered from serum γ-globulins, although the type secreted preferentially (α_1A) is different from the predominant type in serum (α_2M). The ingested γ-globulins are not digested in the neonate, because the gastric contents containing milk buffer are not sufficiently acidic to activate pepsinogen, and colostrum contains trypsin inhibitor to prevent enzymatic breakdown in the intestine. Ingested γ-globulins may protect against diarrheagenic bacteria in the neonatal intestine.

A second important source of protein derives from the intestinal mucosa that is continually sloughed or desquamated into the lumen. Quantitatively, the sloughed protein may amount to half or more of the dietary protein. This fraction is greatly increased in the disease sprue. In sprue, a specific glutamine-containing fraction of wheat protein attacks the microvilli of the intestinal brush border. The microvillous protein is sloughed into the lumen, leaving behind denuded mucosa. In unaffected individuals, an enzyme (gluten hydrolase) normally breaks down the toxic wheat peptide before damage occurs. In sprue patients, protein sources must be sought outside wheat products.

A third important source of protein derives from the many enzymes that are secreted into the gastrointestinal tract. Altogether, the enzymatic protein entering the intestine may exceed the total protein derived from the diet and the sloughed mucosa. Enzymatic protein is probably digested by secreted or bacterial proteases prior to absorption, but the pancreatic enzymes may be absorbed intact by pinocytosis to be re-secreted in the pancreas.

Because the intestinal mucosa is such a leaky epithelium, significant quantities of plasma protein permeate and enter the lumen. This is a normal excretory route for some of the plasma proteins.

Overall, only a very small fraction of protein from all sources escapes small intestinal digestion. This fraction is totally digested by bacteria in the colon. The protein found in the stool derives from excreted bacteria and sloughed colonic mucosa.

Peptide and Amino Acid Absorption

Gastric pepsin and pancreatic peptidases digest protein into primarily di- and tripeptides. The final digestive step to amino acids occurs in the microvilli of the intestinal brush border, where dipeptidases are located. The ultimate digestion to amino acids is the usual but not the necessary preliminary event before absorption. Active mechanisms for the absorption of di- and tripeptides exist in the intestine, and absorption of small peptides can proceed more rapidly than absorption of amino acids. Unlike amino acid absorption, peptide absorption is independent of sodium.

The effectiveness of peptide absorption is illustrated by the lack of malnutrition in patients having genetic disorders of amino acid transport. In cystinuria, the basic amino acids including cystine cannot be absorbed either in the intestine or the kidney. The renal defect is crucial, because cystine in the urine precipitates as calculi (stones) which may obstruct the ureter. The intestinal defect is unnoticeable, because cystine and other basic amino acids can be readily absorbed in the form of di- and tripeptides. In Hartnup's disease, the basic amino acids including tryptophan cannot be absorbed in the intestine. The lack of tryptophan leads to reduced synthesis of nicotinamide and signs of pellagra, but protein malnutrition is absent due to adequate transport of peptides. Bacteria in the colon, however, metabolize the unabsorbed basic amino acids to toxic amines which cause temporary neurological symptoms upon their absorption. In blue diaper syndrome, there is specific lack of tryptophan absorption. In the colon, the unabsorbed tryptophan is converted by bacteria to the blue dye, indigotin, which is absorbed and excreted in the urine. Unabsorbed tryptophan also facilitates calcium absorption, leading to hypercalcemia.

Amino acids are absorbed with apparent specificity from the intestinal lumen. Except for D-methionine, only the naturally occurring levorotatory amino acids are actively transported. Active transport of all amino acids is dependent upon luminal sodium. The carriers for amino acids require simultaneous attachment of sodium ions in order to facilitate entry of both species into the intestinal mucosal cell. The subsequent active extrusion of sodium ions at the basolateral membrane maintains the sodium gradient across the microvillous membrane and thereby maintains the facilitated entry of amino acids. As amino acids accumulate in the mucosal cell, their concentration rises and they passively diffuse across the basolateral membrane into the portal blood. In all steps, coupled sodium and amino acid transport is exactly analogous to coupled sodium and monosaccharide transport in the intestine. Amino acids compete with monosaccharides for intestinal absorption when the metabolic energy for sodium transport becomes limiting.

Several carriers appear to be involved in intestinal amino acid absorption, and attempts have been made to classify the transport mechanisms. Although the classes are notably overlapping in many regards, they do

Table 14–1
Classification of amino acids that are transported across intestinal mucosa

1. Basic	2. Neutral
†lysine	glycine*
hydroxylysine	alanine
arginine	†valine*
histidine*	†leucine*
†tryptophan*	†isoleucine*
cysteine	serine
cystine*	†threonine
†phenylalanine	†methionine
tyrosine	histidine*
	cystine*
	proline*
	hydroxyproline*
3. Acidic	**4. Imino**
glutamic	proline*
aspartic	hydroxyproline*

* Amino acids that fall into more than one class on the basis of competitive or genetic studies.
† Essential amino acids.

suggest at least four different amino acid transport mechanisms: 1) basic or di-amino, mono-carboxylic acids, 2) neutral or mono-amino, mono-carboxylic acids, 3) acidic or mono-amino, di-carboxylic acids, and 4) imino acids. The individual members of these classes are listed in Table 14.1. Basic amino acids are actively absorbed from the intestinal lumen at low concentrations (less than 3 mM). In contrast, neutral amino acids are only absorbed at relatively high concentrations (over 20 mM). Acidic amino and imino acids consist of but two members each. The acidic amino acids are the only amino acids to undergo significant metabolism during intestinal absorption. Glutamic and aspartic acids are transaminated to alanine during absorption of a meal. The overlap between groups is indicated by the large number of amino acids that fall into more than one class, indicated by asterisks (*) in Table 14–1. Besides possible common membership in basic, neutral and imino classes, competitive studies indicate other potential classes: a) glycine and proline; b) valine, leucine, isoleucine. Genetic disorders suggest still more classes: c) cystine (Type 2 cystinuria); d) tryptophan (blue diaper syndrome).

Proteins are almost entirely absorbed from the jejunum. No absorption occurs proximal to the ligament of Treitz, where the common bile duct brings in pancreatic peptidases. Pancreatic peptidases act in the intestinal lumen and while adsorbed to the glycocalyx of the mucosa to hydrolyze rapidly the large peptones left after pepsin digestion. Pancreatic enzymatic digestion of peptones to small peptides and amino acids is not limiting to absorption. Absorption of both small peptides and amino acids is also rapid, and so absorption is normally nearly complete at the end of the jejunum even though all the intestine has the potential for amino acid absorption. Absorption of tryptophan is especially fast, and glycine absorption is relatively slow.

Children with the congenital disorder, cystic fibrosis, have a viscous fluid in the acini and ducts of the exocrine pancreas. For this reason almost none of the pancreatic peptidases can be delivered to the intestine of these children, and they suffer from protein malabsorption and malnutrition. Those pancreatic peptidases which are delivered are subsequently adsorbed onto a viscous glycoprotein secretion of hyperplastic intestinal goblet cells. This abnormal secretion forms a thick mucous layer over the mucosa.

Large numbers of children in underdeveloped countries suffer from protein malnutrition due to protein deficiency, a syndrome known as kwashiorkor. The virtual lack of protein in the diet depletes body stores of protein, including those that comprise pancreatic enzymes. As a result, even the capacity to digest protein is lost. Those children are small in stature, mentally retarded, wasted and suffer from ascites.

Proteins in foods vary in their "biological value" according to their content of essential amino acids (†, Table 14–1). Proteins in milk have a high biological value because they have all the essential amino acids. Proteins in soybean, in contrast, have a low biological value because they lack many of the essential amino acids. When eaten together as part of a normal mixed diet, however, proteins of different biological values complement one another in supplying all of the essential amino acids. The essential amino acids are absorbed by the basic and neutral transport mechanisms (Table 14–1).

Summary

The principal sources of protein found in the intestinal lumen are the diet, the sloughed mucosa and the secreted enzymes. Except for immunoglobulin proteins that may be taken up intact in the intestine of the neonate, protein from all sources is normally digested to small peptides and amino acids prior to absorption. Absorption of tri- and dipeptides is rapid and independent of luminal sodium. Absorption of amino acids requires preliminary dipeptidase activity at the brush border and is dependent upon luminal sodium. Peptide and amino acid absorption are active mechanisms. Many different classes of amino acid transport mechanisms have been described. The largest classes are those containing the basic and neutral amino acids. It is these classes which are affected by genetic disorders and which include the essential amino acids. Protein malnutrition is generally not a result of amino acid malabsorption but rather the consequence of a diet deficient in total protein or protein of biological value.

chapter 15

Vitamin Absorption

how intestinal mucosa absorbs vitamins

In addition to the essential amino acids, the intestinal mucosa absorbs certain other essential substances known as vitamins. The vitamins are a heterogeneous group consisting mostly of coenzymes that are required in various metabolic reactions. Like essential amino acids, vitamins are either not synthesized in humans or the rate of synthesis does not meet minimal requirements and so absorption from dietary sources is critical. Vitamins fall into two broad classes, water-soluble and fat-soluble. Water-soluble vitamins are absorbed by simple passive transport, facilitated passive transport and active transport. Fat-soluble vitamins, in contrast, are absorbed exclusively by passive transport.

Water-Soluble Vitamins

Ascorbic acid is absorbed by simple passive transport in the intestine. Absorption continues to fill tissue stores, which can serve the body for several months after the source of ascorbic acid is removed. There is some conversion of ascorbic acid to 2-ascorbic acid sulfate for storage and excretion. Excess ascorbic acid is excreted into the urine, as are all the water-soluble vitamins.

Biotin is presumably absorbed by simple passive transport, although no experimental evidence exists regarding its mode of absorption.

Cyanocobalamin is absorbed by active transport in the ileum. This absorption depends on several factors, the most important of which is the synthesis of gastric intrinsic factor. Gastric intrinsic factor is a glycoprotein that is produced by gastric oxyntic cells. It binds to cyanocobalamin and forms a complex, which is the required substrate for ileal absorption. Besides gastric intrinsic factor, gastric acid, pancreatic bicarbonate, trypsin and Ca^{++} are required for adequate absorption of cyanocobalamin. Gastric acid is required to free cyanocobalamin from ingested proteins that block the formation of a complex with gastric intrinsic factor. Pancreatic bicarbonate and Ca^{++} provide the required conditions for binding of the complex to the ileal receptor. Trypsin presumably guards against breakdown of the complex by ingested proteins that trypsin digests. After attachment to the ileal receptor, the cyanocobalamin is actively transported into the mucosal cell and the gastric intrinsic factor is released back into the lumen for ultimate excretion. From the mucosal cell cyanocobalamin is transferred to plasma proteins (transcobalamins) for delivery to peripheral tissues.

Cyanocobalamin gets its name from the fact that it contains cobalt and cyanide in its core. It is synthesized exclusively by microorganisms, including some human intestinal bacteria. Much more cyanocobalamin is manufactured by bacterial fermentation in the stomachs of ruminants, and the meat products of these animals are the richest dietary sources of this vitamin.

Lack of cyanocobalamin gives rise to pernicious (macrocytic hyperchromic) anemia. Among the many causes of pernicious anemia are a vegetarian diet, gastric resection or bypass, ileal resection or bypass and pancreatic insufficiency. A fish tapeworm that competes with the human host for cyanocobalamin also causes pernicious anemia. In this country, however, the most common cause is an autoimmune reaction with formation of antibodies against gastric intrinsic factor or oxyntic cells in certain genetically predisposed individuals. Such patients are usually achlorhydric. Binding or production of intrinsic factor is largely abolished, cyanocobalamin is inadequately absorbed and pernicious anemia develops. In the treatment of pernicious anemia, cyanocobalamin is administered parenterally, although high doses of this vitamin can also be ingested, because some small fraction of the free vitamin can be absorbed even in patients with this disease.

Folic acid is absorbed by a carrier-mediated mechanism in the intestine. The efficiency of the active transport of this vitamin is such that less than half the ingested folic acid is absorbed. Human bacteria synthesize some folic acid in the intestine, but most derives from the diet. Dietary folic acid is composed of polyglutamic folic acid and pteroylpolyglutamates. Within the brush border, these polymers are split into monomers for absorption into the mucosal cell.

Niacin is mainly absorbed by simple passive transport in the intestine, although the ileum also has a Na^+-dependent, facilitated transport mechanism.

Pantothenic acid and pyridoxine are thought to be absorbed by simple passive transport.

Riboflavin is absorbed by a facilitated transport mechanism. Some riboflavin is circulated in the enterohepatic system.

Thiamine is absorbed by a Na^+-dependent, active transport mechanism thiamine absorption proceeds by simple passive transport. Bacteria in the human intestine synthesize about 25% of the thiamine requirement. Inadequate dietary thiamine gives rise to a deficiency syndrome found almost exclusively among alcoholics in this country. In severe alcoholism, there may be specific impairment of thiamine absorption due to a simultaneous deficiency of folic acid.

A persistent watery diarrhea may interfere with the absorption of water-soluble vitamins and lead to avitaminoses in susceptible individuals (such as infants and malnourished adults). Malabsorption of these vitamins usually occurs in combination with inadequate uptake of fat-soluble vitamins.

A classification of intestinal mechanisms for vitamin absorption is given in Table 15.1. The tendency has been to drop the letter names for the water-soluble vitamins while retaining alphabetical nomenclature for the fat-soluble vitamins (A, D, E, K). Note the variety of transport mechanisms for water-soluble vitamins, in contrast to the passive absorption of the fat-soluble vitamins.

Fat-Soluble Vitamins

In general, the fat-soluble vitamins are absorbed with fats, and absorption of both lipids will be considered in more detail in the following chapter on lipid absorption. In contrast to the water-soluble vitamins, the fat-soluble vitamins can be stored in relatively high concentrations in the liver, and cases of hypervitaminosis A and D have occurred following chronic ingestion of excess vitamins. Whereas the water-soluble vitamins are excreted in the urine, the principal excretory route for fat-soluble vitamins is fecal.

Vitamin A is present as retinyl ester in foods of animal origin and as β-carotene in foods of plant origin. Retinyl ester is hydrolyzed to retinol

Table 15–1
Mechanisms of vitamin absorption in the intestine

SIMPLE PASSIVE TRANSPORT
Ascorbic Acid (Vitamin C)
Biotin
* Niacin
Pantothenic Acid
Pyridoxine
* Thiamine
† Vitamin A
† Vitamin D
† Vitamin E
FACILITATED PASSIVE TRANSPORT
* Folic Acid
* Niacin
Riboflavin
† Vitamin K
ACTIVE TRANSPORT
Cyanocobalamin (Vitamin B_{12})
* Folic Acid
* Thiamine

* Vitamins absorbed by more than one mechanism.
† Fat-soluble vitamins; all other vitamins are water-soluble. Commonly used alternate names appear in parentheses.

at the brush border membrane prior to passive absorption. Inside the mucosal cell, retinol combines mainly with palmitic acid. The retinyl palmitate is incorporated into chylomicra which are absorbed by lymphatics. β-Carotene is partially converted to retinol, and the remainder is incorporated directly into the chylomicra. Mineral oil, which is ingested as a laxative, interferes with the absorption of vitamin A and the other fat-soluble vitamins by competing with the lipid membranes of the mucosa for dissolution of ingested lipids. As a consequence, much of the vitamin A is excreted in the feces along with the mineral oil.

Vitamin D is passively absorbed into intestinal mucosal cells, incorporated into chylomicra and transferred via the lymphatics to the blood. Vitamin D undergoes metabolic conversion first in the liver and subsequently in the kidney. The terminal product is 1,25-dihydroxycholecalciferol, whose synthesis is under parathyroid hormonal control (as was discussed in connection with calcium absorption in the intestine). This active metabolite of vitamin D induces synthesis of Ca^{++}-binding protein in the brush border and stimulates calcium absorption.

Vitamin E is absorbed passively and relatively inefficiently in the intestine and is incorporated into chylomicra for lymphatic transfer.

Vitamin K is unique in being the only vitamin that is supplied almost totally by resident bacteria in the human intestine. Since the gut is sterile at birth, the limited stores of vitamin K in the infant may be depleted before intestinal flora can restore them; a temporary bleeding tendency can result, because vitamin K is required for synthesis of prothrombin and three coagulation factors. Like the other fat-soluble vitamins and lipids, vitamin K is passively absorbed (with carrier mediation) in the intestine, incorporated into chylomicra and transferred to lymph and blood.

There are many disorders which interfere with absorption of dietary fats from the lumen of the gut. Such malabsorption syndromes are also associated with loss of uptake of fat-soluble vitamins. If the disorder is prolonged, as in a chronic disease such as sprue or with long-term administration of certain drugs which interfere with absorption of fat, signs of specific vitamin deficiencies may occur, *e.g.,* decalcification of bone (avitaminosis D) or spontaneous bleeding (avitaminosis K).

Summary

Intestinal absorption of vitamins lacks a single mechanism due to the wide variety of chemical substances identified as vitamins. The water-soluble vitamins are absorbed by simple passive transport, facilitated passive transport and active transport. The fat-soluble vitamins are absorbed by passive transport involving permeation of the brush border lipid bilayer, incorporation into chylomicra and release into lymph. Many of the vitamins are metabolized either in the intestinal mucosa during absorption or in the liver during storage.

chapter 16

Lipid Absorption

how intestinal mucosa absorbs fat-soluble substances

Because lipids are soluble in the lipid bilayer of the intestinal mucosal membranes, lipids passively permeate the brush border and enter the intestinal mucosal cells. In order for lipid absorption to proceed normally, it is necessary for these substances to dissolve in the watery chyme of the intestinal lumen prior to permeation of the brush border. This dissolution requires a critical concentration of conjugated bile salts to emulsify the lipids and incorporate them within the lumen in small particles called micelles. Lipids are packaged after permeation of the brush border, also. Incorporation of resynthesized lipids within intestinal cells in small particles called chylomicra serves the same purpose of dissolving lipids in the watery media of the cytoplasm, and subsequently in the lymph and the blood.

Fat Digestion

The largest fraction of ingested lipid consists of triglycerides of long-chain fatty acids. These triglycerides must be broken down into fatty acids and monoglycerides for incorporation into micelles and for subsequent absorption. This digestion is accomplished by a large amount of lipase secreted in its active form mainly by the pancreas. Stimulation of pancreatic lipase secretion is a response to cholecystokinin released from intestinal mucosa bathed by chyme containing fatty acids. In this positive feedback loop, the products of fat digestion stimulate secretion of the enzyme catalyzing further fat digestion.

Lipase is secreted in large excess by the pancreas, and so the activity of lipase depends on its substrate concentration. Since its substrate is fat found in globules floating in the watery chyme of the intestinal lumen, the size and resulting surface area of the globules becomes limiting to fat digestion. Before emulsification by bile salts, the fat globules have an average diameter of about 1000 Å. After bile salt emulsification, the average diameter is reduced to only about 50 Å. This reduction in particle size amplifies the total surface area for enzyme activity around 20-fold. This is the reason why lipase acts so much more effectively in the presence of bile salts.

The predominant form of the pancreatic enzyme—lipase B—has an electrically negative charge, as do bile salts. Their mutual repulsion is prevented by a polypeptide cofactor found in pancreatic juice, colipase, which complexes lipase to bile salts at the surface of micelles. Colipase also lowers the pH optimum for lipase activity to near the prevailing pH (6 to 7) found in the intestinal lumen. For these reasons, formation of the lipase-colipase-bile salt complex is the rate-limiting step in lipolysis.

Several forms of lipase are secreted by the pancreas as well as by other tissues. Besides lipase B, the pancreas secretes lipase A and nonspecific lipase. Lipase A is stimulated by bile salts and the latter acts on carboxylic and sterol esters. Glands in the pharynx secrete a lipase that initiates fat digestion by breaking down triglycerides to diglycerides. Pharyngeal lipase is swallowed and acts mainly in the stomach. Intestinal lipase acts mainly on short- and medium-chain triglycerides commonly found in dairy products. Milk fats are digested by a lipase produced in the mammary glands. Milk lipase is resistant to gastric acid and is stimulated by primary bile salts.

Lipids other than triglyceride are digested by other pancreatic enzymes. Dietary cholesterol esters are hydrolyzed by pancreatic cholesterol esterase. Phospholipids are digested by the two types of pancreatic phospholipase. They convert lecithin into lysolecithin.

The lipid substrates, their digestive enzymes and cofactors, and their absorbed product are depicted schematically in Figure 16.1. Note that all dietary lipids except for short- and medium-chain triglycerides require bile salts and micelle formation for absorption.

Micelle Formation

The products of pancreatic lipase, phospholipase and cholesterol esterase incorporate into a shell of bile salts to form a particle called the

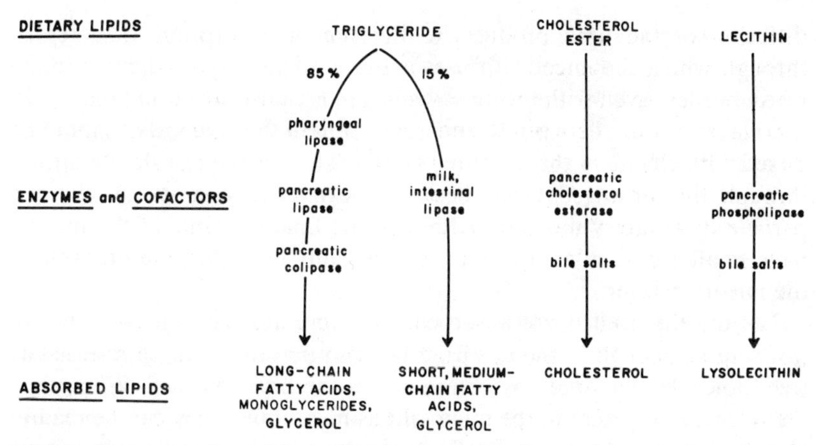

Figure 16.1. Modes of intestinal lipid absorption. Only dietary sources of lipids are shown.

micelle. For a micelle having some 33 bile salt molecules in its outer shell, some 47 fatty acid or monoglyceride molecules will be incorporated in its interior during fat digestion. Another 5 molecules of lysolecithin, 2 molecules of cholesterol and perhaps a molecule of a fat-soluble vitamin will be added to the interior during digestion of phospholipid and cholesterol ester. (Much of the micellar cholesterol also derives from secreted bile.) The bile salt molecules are all oriented with their conjugated, charged ends at the micellar surface, where they interact with the polar solutes and solvents of intestinal chyme. The uncharged, hydrophobic tails of the bile salt molecules are packed in the micelle interior along with the hydrophobic tails of fatty acids, monoglycerides (mostly 2-monoglyceride), lysolecithin (mostly 2-lysolecithin), cholesterol and the fat-soluble vitamins in addition.

The micelle is stabilized by the negative charges carried mainly by conjugated bile salt molecules. When the bile salt concentration is too low to permit this stabilization, micelles fail to form. This critical micellar concentration of conjugated bile salts, about 0.25 mM, is well below the normal average concentration in the intestine after a meal, around 15 mM. Since unconjugated bile salts are less polar than conjugated bile salts, the unconjugated critical micellar concentration is much greater, around 4 mM. Obviously, a considerable safety margin normally insures that micelles form, even when intestinal microflora deconjugate bile salts. With biliary obstruction by gallstones, however, bile salt concentration can fall below the critical micellar concentration.

The micelle is important, because it serves as an essential vehicle for diffusion of the lipid products to the site of absorption. The region through which the micelle diffuses is the unstirred layer adjacent to the brush border. Even with the most vigorous peristalsis this thin (about 0.25 mm) layer of fluid between the microvilli and in the glycocalyx cannot be mixed with chyme in the intestinal lumen. The micelle can slowly diffuse through this unstirred layer, because the micelle is a stable, charged particle in a watery medium. Although the lipid contents of the micelle have smaller molecular sizes, they are virtually insoluble and unstable in the unstirred layer.

The micelle itself is not absorbed. The lipid molecules it contains are just soluble enough in the unstirred layer so that they can be released as free molecules for final absorption across the brush border. When they are released very near to the microvillous membrane, they can be readily absorbed by passive transport. The release from the micelle is a passive consequence of the equilibrium that exists across the micellar surface.

The passage through the unstirred layer is the limiting factor in the absorption of long-chain fatty acids and monoglycerides, cholesterol, lysolecithin and the fat-soluble vitamins. Short- and medium-chain fatty acids and glycerol are sufficiently soluble in the watery chyme that they can freely diffuse through the unstirred layer without the assistance of the micelle. Their absorption is less limited by the unstirred layer.

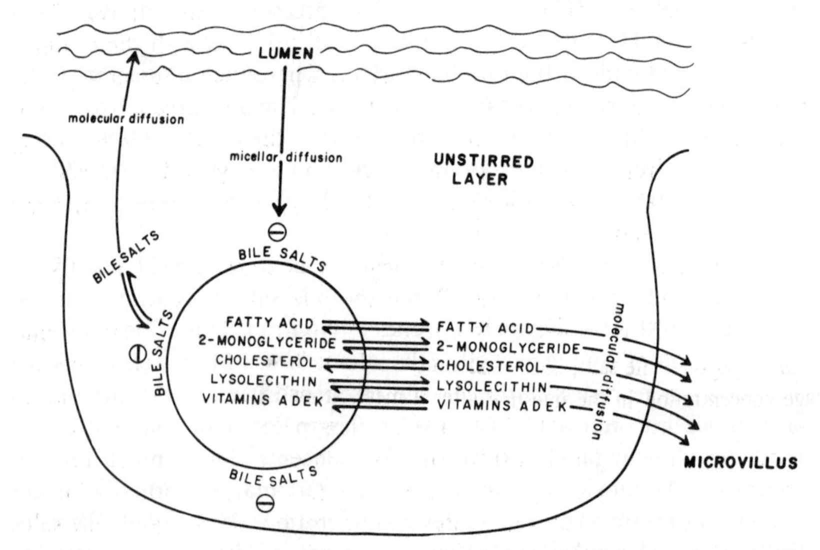

Figure 16.2. Micelle functions in the unstirred layer. Lipid contents of micelle are absorbed by passive transport.

A micelle in the unstirred layer is illustrated in Figure 16.2. Notice that the bile salts do not follow the lipid contents in being absorbed. When all the cholesterol and lysolecithin is absorbed the bile salt shell decomposes, and the bile salts diffuse back through the unstirred layer to the lumen, where they re-form new micelles with lipid constituents. Ultimately, after all the lipid has been absorbed, the conjugated bile salts themselves are actively absorbed in the ileum and into the enterohepatic circulation.

Chylomicron Formation

After the passive absorption of the lipid constituents through the microvillous membrane has taken place, the lipids are processed within the intestinal mucosal cell. This processing consists of resynthesizing triglyceride from absorbed fatty acids and monoglycerides and in packaging triglycerides as well as the other absorbed lipids into particles called chylomicra. Like micelles, chylomicra stabilize lipids in a watery medium, in this case cell water. They have an outer coat of negatively charged β-lipoprotein which is analogous to the bile salts of micelles. Cellular processing and chylomicron formation maintain low cellular concentrations of the absorbed lipids. Thus, their concentration gradients for passive absorption across the microvillous membrane are maintained.

The chylomicron is less than half the size of the micelle, having a diameter of about 20 Å. Chylomicra form in proliferation after a fatty meal. After a proteinaceous meal or during fasting a second type of particle forms in place of the chylomicron. This so-called very low density lipoprotein (VLDL) is smaller than the chylomicron and may be the principal route for dietary cholesterol absorption. Chylomicra and the VLDL diffuse out of the mucosal cell through the basolateral membrane. They accumulate in the lymph of the terminal lacteals, to which chylomicra give a milky appearance. Ultimately, both are delivered via the thoracic duct to the blood of the neck veins.

Short- and medium-chain fatty acids as well as glycerol are passively absorbed directly into the portal blood. Their absorption requires neither micelles nor chylomicra. The glycerol produced in lipolysis is not used in the resynthesis of triglycerides. The resynthesis utilizes endogenously produced glycerol derived from α-glycerophosphate.

Cholesterol absorption presents some interesting special features. In its absorption, it appears to utilize VLDL in place of chylomicra. Dietary cholesterol is additionally sequestered from nondietary cholesterol in mucosal cells. Of the total cholesterol absorbed in the intestine, less than 20% derives from dietary cholesterol ester. The rest derives from desquamated mucosal membranes (less than 20%) and mainly from bile (more than 60%). In addition, the intestinal mucosa synthesizes cholesterol at a rate which is inversely proportional to the rate of absorption. Thus, cholesterol synthesis by intestinal mucosa increases on a low-cholesterol diet. The nondietary and synthesized cholesterol form two pools which attach to high density lipoprotein of the plasma. These lipoproteins are stable in the plasma. Dietary cholesterol, in contrast, forms a separate pool which is attached to low and very low density lipoproteins of the plasma. These lipoproteins are relatively unstable in the plasma and may deposit cholesterol as plaques on the intima of blood vessels. For this reason, a diet high in cholesterol has been associated with athero- and arteriosclerosis.

Steps in the absorption of cholesterol are shown in Figure 16.3. How the three pools of cholesterol are segregated within the mucosal cell is not known.

Lipid Malabsorption

Failure of the intestine to absorb lipids is caused by two types of disease. First, malabsorption of lipid can arise from failure to emulsify or digest intraluminal fat which may be caused by such diseases as pancreatitis, pancreatic carcinoma and cystic fibrosis, in which pancreatic enzymes are not secreted, or by hepatitis and blockage of the common bile duct by gallstones, in which case bile salts are unavailable for micelle formation. Second, malabsorption can follow from loss of normal intestinal mucosa required for absorption itself; such loss can be caused by surgical removal or bypass of small intestine, as part of a reaction to drugs or as a result of a congenital deficiency of an enzyme, as in sprue.

Gluten hydrolase is an intestinal enzyme which breaks down gluten, a glutamine-containing polypeptide found in wheat products. In the absence of this hydrolase, a toxic peptide (gliadin fraction) remains to attack the crypts of Lieberkühn in the intestinal mucosa. The principal effect of this action is the failure of microvilli to form. Lack of microvilli severely retards absorption of lipids, because the resulting mucosal surface area for their passive transport is reduced some 20-fold.

Besides reducing the intestinal surface area for lipid absorption, the toxic peptide in sprue causes lipid malabsorption in other ways. The denuded intestinal mucosal membrane cannot transport Na^+, nutrients and water, and so the normal net absorption of fluid reverses to a net secretion. Secretory flow in turn produces a solvent drag on the diffusion of micelles through the unstirred layer. This serves to further limit lipid

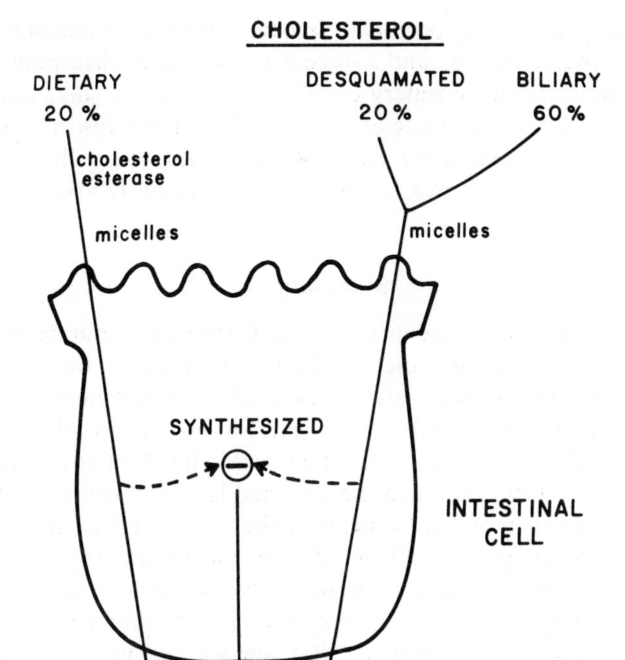

Figure 16.3. Special features of cholesterol absorption. The amount of cholesterol absorbed from the lumen exerts negative feedback on the amount of cholesterol synthesized in the cell, indicated by *dashed arrows* and ⊖. *VLDL*, very low density lipoprotein; *HDL*, high density lipoprotein.

absorption. Failure to absorb nutrients leads to the establishment of an osmotic gradient favoring secretion, because digestive enzymes and intestinal microflora continue to break down food polymers into their osmotically more active monomers which are not absorbed. The solvent drag effect to slow down lipid absorption is thus further accentuated. The abnormal mucosal membrane in sprue also directly retards the passive entry of lipids. The synthetic enzymes for the reconstitution of triglyceride in the mucosal cell are inactive in sprue. Hence this most important lipid cannot be absorbed into the lymph at all. Intestinal bacteria proliferate in the nutrient-rich intestinal lumen of the sprue patient. They deconjugate bile salts in the lumen, thereby increasing the critical micellar concentration and decreasing lipid absorption. In excess, the deconjugated bile salts irritate the colonic mucosa and reduce Na^+ and water absorption. Thus, sprue is characterized by watery diarrhea as well as by steatorrhea, the appearance of fat in the feces as a result of lipid malabsorption. After many years of inflammatory response to gliadin, the villi disappear, further reducing surface area for absorption of all nutrients, electrolytes and water.

Summary

Absorption of lipids in the intestine is entirely passive but nevertheless complicated. The common dietary lipids are prepared for absorption through digestion by pancreatic lipase and other enzymes. Digestive activity is greatly enhanced by the emulsifying action of bile salts. Digestive products are incorporated into micelles that are soluble in chyme. Micelles diffuse through the unstirred layer to the very surface of the microvillous membrane. Contents of the micelle diffuse to the membrane and passively permeate it. Within the mucosal cell, triglycerides are resynthesized from absorbed fatty acids and endogenous glycerol. The resynthesized triglycerides along with most of the other absorbed lipids are incorporated into chylomicra that are soluble in cytoplasm and extracellular fluid. Chylomicra passively permeate the basolateral membrane into the lymph of the lacteals. Lipids in chylomicra ultimately are discharged into the systemic circulation. Malabsorption of lipids results from inadequate lipase, bile salts or mucosal membrane surface area.

chapter 17

Overview of the Gastrointestinal System

a summary of gastrointestinal functions considered in this book

The gastrointestinal tract is organized essentially as a long tubular structure into whose hollow interior secretions pour from the glands lining the lumen and from nearby solid glandular organs. In the proximal part of the system solid foods and fluid are ingested, chewed, macerated, diluted and digested into simpler molecular forms prior to their absorption from the lumen of the small intestine. These numerous actions upon eaten materials take place successively in the mouth, pharynx, esophagus, stomach and upper small intestine, and require mechanical propulsion of the nutrients from hollow organ to hollow organ. In the upper small intestine particularly, several well-defined transport processes are involved in moving the nutrient molecules from the lumen across the lining membrane into the mucosal cells and subsequently from the cells into the blood. These absorptive processes depend upon the mucosal structure, metabolism and circulation. Beside absorbing materials from the lumen, the mucosal lining of the gastrointestinal tract is also capable of secreting juices into the lumen. Materials which are not absorbed from the gut are moved along to the colon and eliminated as feces.

There are several factors which influence the aforementioned processes. These include metabolic and electrophysiological events taking place in mucosal and smooth muscle cells, and hemodynamic events in the gastrointestinal circulation. In addition, gastrointestinal functions are regulated extensively by both the nervous system and circulating hormones. The gastrointestinal tract provides both the cells of origin and the target cells for these hormones. The gastrointestinal tract is also the site of an extensive network of intrinsic nerves and is subject to the actions of extrinsic autonomic nerves.

In the gastrointestinal mucosa glucose and fatty acids are oxidized to provide the chemical energy upon which both secretion and absorption depend. Other metabolic processes generate NADH and subsequently ATP; dephosphorylation of the latter high energy material by specific ATPases permits gastrointestinal membranes to transport a variety of ions and organic materials against electrochemical gradients in the processes of active secretion and absorption. There is a high rate of turnover of mucosal lining cells with a half-life measured in less than a few days. Sloughed mucosal cells contain enzymes which are active for a time in the lumenal digestive processes. Growth and differentiation of these mucosal cells are influenced by the hormone gastrin.

The movement of dissolved substances across the gastrointestinal mucosa proceeds as one of two types of membrane transport. Passive transport consists of the movement of materials in solution across a membrane in response to existing electrochemical gradients which drive the solutes and to osmotic gradients which attract water across membranes. Active transport permits the movement of materials against their electrochemical gradients but requires the expenditure of energy by the membrane. Materials which are actively transported must attach to a carrier molecule on one side of the membrane and be moved to the other side of the membrane with the expenditure of energy before being released.

The gastrointestinal circulation supports both absorptive and secretory activity in the system. Adequate blood flow to gastrointestinal organs is influenced by the overall state of the circulation of the body, the autonomic nervous system, circulating neurohumoral substances, tissue metabolites and intrinsic vascular characteristics (autoregulation, escape, redistribution, and the countercurrent multiplier). In general, in glandular organs (such as the salivary glands, stomach and pancreas) an increase in secretory rate is accompanied by an increase in metabolism with release of dilator metabolites and an increase in blood flow to the gland.

Gastrointestinal hormones originate in the APUD cells of the antral or small intestinal mucosa. These hormones are polypeptides which belong to a few families of compounds. Events relating to the processes of feeding, digestion and absorption appear to stimulate release of these agents. Among the numerous hormones which have been identified to date, there are currently four whose physiological importance appears established; these are gastrin, secretin, cholecystokinin (CCK) and gastric inhibitory peptide (GIP). Gastrin stimulates secretion of acid by the stomach and is a growth hormone for the gastrointestinal mucosa. Secretin stimulates secretion of an alkaline pancreatic juice. CCK stimulates pancreatic secretion of digestive enzymes and causes emptying of the gallbladder. GIP is insulinotropic and inhibits both secretion and emptying of the stomach.

The neural components of the gastrointestinal system consist of three functionally different kinds of nerves: 1) extrinsic sympathetic and parasympathetic efferent fibers; 2) afferent nerves from the gastrointestinal organs; and 3) intrinsic nerves which exist in plexuses within the walls of the gastrointestinal organs. Neural activity influences glandular secretion, motility and release of gastrointestinal hormones. Afferent nerves from the gastrointestinal tract convey the sense of pain and inform us when we are hungry. Parasympathetic efferent nerves stimulate salivation, gastric secretion of acid and pepsin, pancreatic secretion of enzymes, the release of gastrin into the blood and changes in motor activity in the smooth muscle walls of the hollow organs.

The walls of the hollow organs of the gastrointestinal tract contain a large amount of smooth muscle, whose movements are termed "motility." Such movements can be propulsive or can enhance the mixing of the contents of a hollow organ. In the absence of food in the gastrointestinal system there is a continual low level of motor activity which is augmented at mealtime by central neural messages, by intrinsic nervous responses to food in the tract and by hormones. Swallowing is a complex reflex which propels chewed food from the mouth into the esophagus. The body of the esophagus exhibits a type of movement called "peristalsis" which propels food and fluid along the length of the organ. At the junction between the esophagus and the stomach is the lower esophageal sphincter which is usually constricted, except at times when food is approaching. The upper portion of the stomach behaves as a storage site for swallowed food and fluid whereas the lower part of the stomach exhibits peristaltic motions which slowly propel dissolved food and fluid into the upper small intestine. Factors which appear to regulate emptying of the stomach include characteristics of the food (osmolality, fat and protein content), the level of motor activity in the stomach and duodenum, the tone of the pyloric sphincter, extrinsic and intrinsic parasympathetic nerves and the gastrointestinal hormones (gastrin, secretin, CCK and GIP). In the small intestine there is both propulsive motor action and mixing of the chyme to facilitate digestion and absorption of nutrients. In the colon there is limited propulsive motor activity, since the contents usually remain in that organ for some days before expulsion.

Secretions of the salivary glands participate in the process of digestion of food. Saliva contains an enzyme, α-amylase, which breaks down starch into smaller molecules. Saliva also lubricates and macerates chewed food.

Oxyntic cells of the gastric mucosa secrete solutions of hydrochloric acid in remarkably high concentrations into the lumen of the stomach. The process is energy-consuming and requires initial hydrolysis of water into hydrogen and hydroxyl ions. Subsequently the membranes of these same cells actively secrete hydrogen and chloride ions in a juice which is essentially isosmotic HCl. The residual hydroxyl ion inside the oxyntic cell is combined with CO_2 to form bicarbonate which is then transported into the blood. A highly acidic gastric juice is required to activate pepsin, an enzyme active in the digestion of protein.

The pancreatic juice is essential for proper digestion of food. The juice is highly alkaline and contains a wide array of enzymes which participate actively in the breakdown of carbohydrates, fat and protein ingested in our food. This action results in the presentation of simpler molecules to the small intestine for absorption. Pancreatic production of the alkaline component of the juice is stimulated by the hormone secretin, and the release of enzymes from the pancreas is stimulated by vagal input and by the hormone CCK.

In the liver primary bile acids are synthesized from cholesterol and are conjugated to glycine or taurine before being secreted into the canaliculi. The secretion of the liver is called bile and consists of an alkaline solution containing much bicarbonate and bile salts. Bile is stored in the gallbladder where the organic components are concentrated as a result of the active absorption of sodium by the mucosa of the gallbladder. At mealtime the bile is extruded into the duodenum where the bile salts emulsify fat. After participating in the digestion of fat in the gut, bile salts are either deconjugated by bacteria and passively absorbed in the gut or are propelled to the ileum where they are actively absorbed in the conjugated state. Absorbed bile salts are carried through the portal circulation to the liver for resecretion. The pathway taken by bile salts from the liver through the biliary tree to the intestine and reabsorption into the portal circulation before being returned to the liver constitutes the enterohepatic circulation.

The major process involved in the net absorption of fluid and electrolytes from the lumen of the gut is the active transport of sodium which carries along chloride and water passively. This movement of sodium is also closely linked to the active transport of glucose and amino acids. Because of the high order of efficiency of sodium transport, nearly all of the daily fluid entering the small intestine is absorbed. Other electrolytes of importance which are absorbed from the small intestine include calcium and iron.

Small intestinal absorption of carbohydrates depends upon prior digestion of the food, the integrity of the mucosa, the effectiveness of the transport mechanisms for carbohydrates, intestinal transit time and mechanical mixing of the chyme by the smooth muscle wall of the gut. In the average adult diet about 250 g of carbohydrates are consumed, for the most part as polysaccharides or disaccharides. In these molecular forms carbohydrates are not absorbed effectively from the lumen of the gut. Several enzymes are responsible for digestion of these complex molecules into monosaccharides, including salivary and pancreatic α-amylases, and the disaccharidases located on the brush border of the villi. About 80% of this enzymatic breakdown of carbohydrates is in the form of glucose which is absorbed largely from the lumen of the gut by a sodium-dependent active transport mechanism. The carrier for glucose also binds galactose. Fructose is absorbed by a passive facilitated transport process.

The major sources of protein in the intestinal lumen are the diet, sloughed mucosa and enzymes. Within the lumen enzymatic digestion breaks down the protein into amino acids and small peptides. Tripeptides and dipeptides are actively transported by the intestinal mucosa. There are multiple separate active transport mechanisms for amino acids and such transport is dependent upon the presence of sodium.

There are a variety of independent mechanisms involved in the absorption of vitamins. Water-soluble vitamins are absorbed by simple passive transport, facilitated passive transport and active transport. Fat-soluble vitamins are absorbed by passive transport whereby these vitamins readily penetrate the brush border membrane and are incorporated into chylomicra within the mucosal cells. Subsequently the chylomicra are transported from the cell into the lymphatics.

Absorption of fats from the lumen of the small intestine is a passive transport process. Digestion of dietary fat in the lumen is mediated primarily by pancreatic lipase and related enzymes but is also dependent upon emulsification of the dietary lipid by bile salts. As digestion proceeds the simpler molecules are incorporated into dissolved molecular accumulations, called micelles. Micelles diffuse through the unstirred layer of chyme adjacent to the mucosa and reach the membrane of the microvilli.

The micelle disintegrates here and its lipid components diffuse through the membrane. Inside the mucosal cell, some triglycerides are resynthesized from the simpler molecules (fatty acids, monoglycerides and glycerol). A variety of lipid molecules are incorporated inside the cell into chylomicra which are soluble in the cytoplasm. Chylomicra penetrate the basolateral membrane, enter the lymphatics and are subsequently carried to the blood.

Index